MODELING MOBILE-SOURCE EMISSIONS

Committee to Review EPA's Mobile Source Emissions Factor (MOBILE) Model
Board on Environmental Studies and Toxicology
Commission on Geosciences, Environment, and Resources
Transportation Research Board
National Research Council

NATIONAL ACADEMY PRESS
Washington, D.C.

NATIONAL ACADEMY PRESS 2101 Constitution Ave., N.W. Washington, D.C. 20418

NOTICE: The project that is the subject of this report was approved by the Governing Board of the National Research Council, whose members are drawn from the councils of the National Academy of Sciences, the National Academy of Engineering, and the Institute of Medicine. The members of the committee responsible for the report were chosen for their special competences and with regard for appropriate balance.
This project was supported by Grant No. X825773–01–0 between the National Academy of Sciences and the U.S. Environmental Protection Agency and the U.S. Department of Transportation. Any opinions, findings, conclusions, or recommendations expressed in this publication are those of the author(s) and do not necessarily reflect the view of the organizations or agencies that provided support for this project.

International Standard Book Number 0-309-07086-0
Additional copies of this report are available from: National Academy Press 2101 Constitution Ave., NW Box 285 Washington, DC 20055 800–624–6242 202–334–3313 (in the Washington metropolitan area) http://www.nap.edu

Printed in the United States of America

THE NATIONAL ACADEMIES

National Academy of Sciences
National Academy of Engineering
Institute of Medicine
National Research Council

The **National Academy of Sciences** is a private, nonprofit, self-perpetuating society of distinguished scholars engaged in scientific and engineering research, dedicated to the furtherance of science and technology and to their use for the general welfare. Upon the authority of the charter granted to it by the Congress in 1863, the Academy has a mandate that requires it to advise the federal government on scientific and technical matters. Dr. Bruce M. Alberts is president of the National Academy of Sciences.

The **National Academy of Engineering** was established in 1964, under the charter of the National Academy of Sciences, as a parallel organization of outstanding engineers. It is autonomous in its administration and in the selection of its members, sharing with the National Academy of Sciences the responsibility for advising the federal government. The National Academy of Engineering also sponsors engineering programs aimed at meeting national needs, encourages education and research, and recognizes the superior achievements of engineers. Dr. William A. Wulf is president of the National Academy of Engineering.

The **Institute of Medicine** was established in 1970 by the National Academy of Sciences to secure the services of eminent members of appropriate professions in the examination of policy matters pertaining to the health of the public. The Institute acts under the responsibility given to the National Academy of Sciences by its congressional charter to be an adviser to the federal government and, upon its own initiative, to identify issues of medical care, research, and education. Dr. Kenneth I. Shine is president of the Institute of Medicine.

The **National Research Council** was organized by the National Academy of Sciences in 1916 to associate the broad community of science and technology with the Academy's purposes of furthering knowledge and advising the federal government. Functioning in accordance with general policies determined by the Academy, the Council has become the principal operating agency of both the National Academy of Sciences and the National Academy of Engineering in providing services to the government, the public, and the scientific and engineering communities. The Council is administered jointly by both Academies and the Institute of Medicine. Dr. Bruce M. Alberts and Dr. William A. Wulf are chairman and vice chairman, respectively, of the National Research Council.

www.national-academies.org

COMMITTEE TO REVIEW EPA's MOBILE SOURCE EMISSIONS FACTOR (MOBILE) MODEL

ARMISTEAD G.RUSSELL *(Chair),* Georgia Institute of Technology, Atlanta, Georgia
JOHN C.BAILAR III, University of Chicago, Chicago, Illinois
MATTHEW J.BARTH, University of California, Riverside, California
LAURENCE S.CARETTO, California State University, Northridge, California
CARLETON J.HOWARD, National Oceanic and Atmospheric Administration, Boulder, Colorado
JOHN H.JOHNSON, Michigan Technological University, Houghton, Michigan
JOHN F.KOWALCZYK, (retired) Oregon Department of Environmental Quality, Portland, Oregon
ALAN C.LLOYD, California Air Resources Board, Sacramento, California
MICHAEL R.MORRIS, North Central Texas Council of Governments, Arlington, Texas
ALISON K.POLLACK, ENVIRON International Corporation, Novato, California
ROBERT F.SAWYER, University of California, Berkeley, California

Staff

RAYMOND WASSEL, Senior Program Director for Environmental Sciences and Engineering
K.JOHN HOLMES, Project Director
NANCY HUMPHREY, Senior Staff Office
ROBERT CROSSGROVE, Editor
MIRSADA KARALIC-LONCAREVIC, Information Specialist
PAMELA FRIEDMAN, Project Assistant
CHRISTINE PHILLIPS, Project Assistant

Sponsors

U.S. ENVIRONMENTAL PROTECTION AGENCY
U.S. DEPARTMENT OF TRANSPORTATION

BOARD ON ENVIRONMENTAL STUDIES AND TOXICOLOGY

TRANSPORTATION RESEARCH BOARD 2000 EXECUTIVE COMMITTEE

COMMISSION ON GEOSCIENCES, ENVIRONMENT, AND RESOURCES

OTHER REPORTS OF THE BOARD ON ENVIRONMENTAL STUDIES AND TOXICOLOGY

Strengthening Science at the U.S. Environmental Protection Agency: Research Management and Peer Review Practice (2000)

Scientific Frontiers in Developmental Toxicology and Risk Assessment (2000)

Copper in Drinking Water (2000)

Ecological Indicators for the Nation (2000)

Waste Incineration and Public Health (1999)

Hormonally Active Agents in the Environment (1999)

Research Priorities for Airborne Particulate Matter: I. Immediate Priorities and a Long-Range Research Portfolio (1998); II. Evaluating Research Progress and Updating the Portfolio (1999)

Ozone-Forming Potential of Reformulated Gasoline (1999)

Risk-Based Waste Classification in California (1999)

Arsenic in Drinking Water (1999)

Brucellosis in the Greater Yellowstone Area (1998)

The National Research Council's Committee on Toxicology: The First 50 Years (1997)

Toxicologic Assessment of the Army's Zinc Cadmium Sulfide Dispersion Tests (1997)

Carcinogens and Anticarcinogens in the Human Diet (1996)

Upstream: Salmon and Society in the Pacific Northwest (1996)

Science and the Endangered Species Act (1995)

Wetlands: Characteristics and Boundaries (1995)

Biologic Markers (5 reports, 1989–1995)

Review of EPA's Environmental Monitoring and Assessment Program (3 reports, 1994–1995)

Science and Judgment in Risk Assessment (1994)

Ranking Hazardous Waste Sites for Remedial Action (1994)

Pesticides in the Diets of Infants and Children (1993)

Issues in Risk Assessment (1993)

Setting Priorities for Land Conservation (1993)

Protecting Visibility in National Parks and Wilderness Areas (1993)

Dolphins and the Tuna Industry (1992)

Hazardous Materials on the Public Lands (1992)

Science and the National Parks (1992)

Animals as Sentinels of Environmental Health Hazards (1991)

Assessment of the U.S. Outer Continental Shelf Environmental Studies Program, Volumes I-IV (1991–1993)

Human Exposure Assessment for Airborne Pollutants (1991)

Monitoring Human Tissues for Toxic Substances (1991)

Rethinking the Ozone Problem in Urban and Regional Air Pollution (1991)

Decline of the Sea Turtles (1990)

Preface

MOTOR VEHICLES are the major sources of air pollutant emissions in U.S. cities. The constituents of those emissions can be harmful to human health and the environment, and also are responsible for the formation of other harmful compounds, such as ozone and particulate matter. Effectively managing air pollution problems depend critically on having accurate automotive emissions estimates. In the United States (outside of California), the Mobile Source Emissions Factor (MOBILE) Model has been at the heart of this process. MOBILE develops emissions factors that, along with information on vehicle activity, are used to estimate emissions inventories for on-road mobile sources. Further, MOBILE is used to adjust those emissions factors to account for the impact of controls, and to forecast how emissions will change in the future and the effectiveness of control programs.

Studies of the MOBILE model suggest that its ability to accurately assess the effectiveness of various very expensive programs, such as the oxygenated fuels program and inspection and maintenance, is poor. Because of the model's importance in assessing air-quality control programs and because of concerns about weaknesses in the accuracy and reliability of the model, Congress asked the National Academy of Sciences to review the MOBILE model. The National Research Council's Committee to Review EPA's Mobile-Source Emissions Factor Model was formed in response to that request. The task for the committee was to review and evaluate the MOBILE model. Specifically, it was to consider the adequacy of the model's input data, assumptions, structure, and results for mobile-source emissions estimation, and recommend ways to improve the reliability of the model.

Many individuals assisted the committee by providing information on the sources of emissions and emissions modeling techniques addressed in this report. I gratefully acknowledge Mark Carlock, California Air Resources Board; Thomas Darlington, Air Improvement Resources Inc.; Axel Friedrich, German Federal

Environmental Agency; Eric Fujita, Desert Research Institute; Richard Gibbs, New York Department of Environmental Conservation; Randall Guensler, Georgia Institute of Technology; Jose Luis Jimenez, Massachusetts Institute of Technology; Charles Schleyer, Mobil Oil; Joel Schwartz, California Inspection and Maintenance Review Committee; and Thomas Wenzel, Ernest Orlando Lawrence Berkeley National Laboratory. I also thank Philip Lorang, Lois Platte, and John White from the U.S. Environmental Protection Agency and Richard Schoeneberg from the U.S. Department of Transportation Federal Highway Administration for providing information to the committee.

This report has been reviewed in draft form by individuals chosen for their diverse perspectives and technical expertise in accordance with procedures approved by the NRC's Report Review Committee. The purpose of this independent review is to provide candid and critical comments that assist the NRC in making the published report as sound as possible and to ensure that the report meets institutional standards for objectivity, evidence, and responsiveness to the study charge. The review comments and draft manuscript remain confidential to protect the integrity of the deliberative process. The committee wishes to thank the following individuals for their participation in the review of this report: Robert Dulla, Sierra Research, Inc.; Robert Frosch, Harvard University; Eric Fujita, Desert Research Institute; Thomas Graedel, Yale University; Randall Guensler, Georgia Institute of Technology; Winston Harrington, Resources for the Future; David Lax, American Petroleum Institute; Frederick Lurmann, Sonoma Technology, Inc.; John McTague, Ford Motor Company (retired); and Deborah Niemeier, University of California, Davis.

The individuals listed above have provided many constructive comments and suggestions. It must be emphasized, however, that responsibility for the final content of this report rests entirely with the authoring committee and the NRC.

I am also grateful for the assistance of the NRC staff in the preparation of this report. K.John Holmes was key in preparing this report in his role as project director. The committee also acknowledges Raymond Wassel, senior program director of environmental sciences and engineering in the Board on Environmental Studies and Toxicology. We also thank the other staff members contributing to this report, including Robert Hamilton, executive director of the Commission on Geosciences, Environment, and Resources; James Reisa, director of the Board on Environmental Studies and Toxicology; Robert Crossgrove, editor; Nancy Humphrey, senior program director with the Transportation Research Council; and Pamela Friedman, Christine Phillips, Tracie Holby, and Ruth Danoff, project assistants.

Finally, I would like to thank all the members of the committee for their expertise and dedicated effort throughout the study.

Armistead G.Russell, Ph.D.
Chair, Committee to Review EPA's Mobile Source Emissions Factor (MOBILE) Model

Contents

Modeling Mobile-Source Emissions

Executive Summary

HE MOBILE SOURCE EMISSIONS FACTOR (MOBILE) model is a computer model developed by the U.S. Environmental Protection Agency (EPA) for estimating emissions from on-road motor vehicles. MOBILE is used in air-quality planning and regulation for estimating emissions of carbon monoxide (CO), volatile organic compounds (VOCs), and nitrogen oxides (NO_x) and for predicting the effects of emissions-reduction programs.[1] Because of its important role in air-quality management, the accuracy of MOBILE is critical. Possible consequences of inaccurately characterizing motor-vehicle emissions include the implementation of insufficient controls that endanger the environment and public health or the implementation of ineffective policies that impose excessive control costs. Billions of dollars per year in transportation funding are linked to air-quality attainment plans, which rely on estimates of mobile-source emissions. Transportation infrastructure decisions are also affected by emissions estimates from MOBILE.

[1]The MOBILE model estimates emissions factors in grams per mile from information on the emissions characteristics of the vehicle fleet. Combining emissions factors with estimates of vehicle miles traveled produces emissions estimates. Particulate matter and air toxics (hazardous air pollutants) are estimated by the PART5 and MOBTOX models, respectively. Because these models are closely tied to MOBILE, they are discussed in the same context as MOBILE in this report. A new version of this model, MOBILE6, is being developed and is expected to be released in the summer of 2000. MOBILE5b is the version currently in use.

There are numerous complexities involved in estimating and predicting mobile-source emissions from the on-road fleet. The fleet is made up of vehicles with a wide variety of emissions characteristics due to differences in condition, type, and age of vehicles; performance of the emissions-control systems; and fuel composition. Emissions also are affected by local factors, such as meteorological conditions, traffic patterns, and travel activity. Developing predictions of future emissions requires projections for all of those characteristics. Finally, there is the issue of the spatial and temporal resolution required for the estimates. For example, estimating the effect of a regional emissions inspection and maintenance program would not require emissions estimates to be resolved at the level of detail needed for estimating the effect of an effort to coordinate traffic signals along a major corridor.

Since its release more than 20 years ago, MOBILE has been used increasingly in regulatory applications. MOBILE is now the central tool used by environmental and transportation agencies to estimate on-road mobile-source emissions and to assess national, state, and local air-quality programs for controlling such emissions. The wide range of mobile-source emissions-control programs assessed with MOBILE varies from national vehicle emissions and fuel standards to local travel demand and congestion mitigation measures.

This wide range of applications has placed greater demands on the model for accuracy and has opened it up to intense scrutiny. Questions have been raised about MOBILE's capability to evaluate reliably the impacts of air-quality-improvement initiatives, such as the vehicle-emissions inspection and maintenance programs and the use of oxygenates in wintertime gasoline blends. Previous and current versions of the model have been criticized for their lack of adequate documentation on underlying methodologies and data. There has also been criticism by the U.S. General Accounting Office that EPA's policy on peer review had not been fully followed during the development of current and past versions of MOBILE.

In response to a request from Congress, the National Research Council established the Committee to Review EPA's Mobile Source Emissions Factor (MOBILE) Model in October 1998. The committee was charged to evaluate MOBILE and to develop recommendations for improving the model. The full charge to the committee is given in Chapter 1.

In carrying out its charge, the committee reviewed the structure and performance of the MOBILE model and considered ways to improve the model. The committee considered MOBILE in the context of its various applications, which include estimating on-road mobile-source emissions and predicting the efficacy of emissions-control strategies. In addition, the committee surveyed developments in other areas of mobile-source emissions modeling.

Although the necessary studies for a quantitative assessment of MOBILE's overall accuracy have not been conducted, some studies show that model estimates of emissions and effectiveness of control strategies are significantly different from those occurring in the real world. For example, studies relying on ambient observations and other field measurements indicate that MOBILE is substantially underestimating mobile-source VOC emissions. Such differences lead to questions about the level of accuracy needed for the intended applications of MOBILE. In response, additional testing of vehicles, rigorous evaluation of emissions estimates, model validation, and sensitivity analysis of the model are needed. These are resource-intensive efforts, but they are necessary for providing sufficient confidence in the emissions estimates from the model.

The committee commends EPA for its response to a number of previous criticisms of the model as it continues to revise MOBILE. In particular, the documentation for MOBILE6, EPA's new version of the model to be released in the near future, is greatly superior to that prepared for previous versions of the model. EPA has attempted to "open up" the model development to users, stakeholders, and other interested parties by holding public workshops, providing detailed documentation, and asking for feedback. That is a positive step by EPA. MOBILE's capabilities have expanded over time, and many improvements have been incorporated into MOBILE6.

MOBILE-SOURCE EMISSIONS MODELING RECOMMENDATIONS

Development of a Toolkit of Models

Finding

Since its development in 1978, MOBILE has evolved from a tool for estimating regional emissions inventories to such uses as determining the conformity of transportation projects with requirements of State Implementation Plans and assessing the emissions impacts of transportation-control measures. The further the model's application deviates from its original purpose of estimating aggregate regional emissions, the more difficult it becomes to verify the accuracy of its predictions and, as a result, the less appropriate it becomes for air-quality management.

That finding is not a basis for the elimination of MOBILE. Instead, it indicates that an upgraded MOBILE model should be included with other emissions models in a toolkit devised to tackle the wide range of current applications. Such a modeling toolkit would provide flexibility for developing mobile-source emissions estimates in response to the wide variety of emissions-control strategies.

Recommendation

Because no single mobile-source emissions model is appropriate for all applications, the committee recommends development of a toolkit of models that includes the following:

- an *aggregated regional emissions-factor modeling component (i.e., the updated MOBILE model)* for estimating emissions using aggregate vehicle-activity data;
- a *mesoscale emissions modeling component* that integrates detailed transportation and emissions components to estimate regional and subregional (corridor) emissions through the coupling of vehicle operating conditions with appropriate emissions factors; and
- a *microscale instantaneous emissions modeling component* that uses instantaneous operating conditions of individual vehicles to estimate continuous vehicle emissions and that can be used for a variety of applications, including generating emissions factors for microscale traffic-simulation models, mesoscale emissions models, traffic data sets, and dispersion models.

To allow for a nested progression of models, the models in the toolkit should be designed and maintained to be consistent, despite differences in the level of detail in inputs (e.g., variations in roadway network detail and meteorological variables) and outputs (e.g., variations in emissions inputs over space and time). Consistency requires that the different modeling components in the toolkit be based on a consistent data set to the maximum extent possible. Consistency also requires that the models in the toolkit predict similar emissions for spatial scales that overlap.

EPA will need to develop a procedure for approving new components of the toolkit for use by states and regions. Such a procedure will require development of guidance documentation describing the appropriate use of each component in the toolkit and technical documentation describing modeling methodologies and data sources. In addition, peer review and model validation must be part of the foundation for new model adoption. Validation efforts for all new modeling methods should be conducted with vehicles and test conditions not reflected in the data used to develop the model and undertaken at the scale (or scales) for which a model is designed.

Model Evaluation

Finding

Model validation and evaluation have not been addressed adequately by EPA during MOBILE's development. MOBILE's *predictions* of the bene fits of air-quality programs (e.g., vehicle emissions inspection and maintenance, oxygenated fuels, and reformulated gasoline) are often taken as *measurements* of the benefits of these programs. Confidence in the model has been undermined when large discrepancies have been observed between the model's predictions and field measurements. Proper testing and evaluation would improve the accuracy of mobile-source emissions modeling in estimating emissions, estimating the effects of emissions on human health and the environment, and estimating the effectiveness of control strategies.

Recommendation

Enhanced model evaluation studies should begin immediately and continue throughout the long-term evolution and development of mobile-source emissions models. These studies should be done with oversight and guidance from a reviewing body such as the EPA Science Advisory Board that includes users and technical experts. They should be undertaken in tandem with the uncertainty studies suggested in the following recommendation.

Evaluation studies should be conducted to identify and reduce disparities between model-predicted emissions and measured data on emissions and air quality. These studies should also focus on reducing the differences between the model-predicted changes in emissions resulting from programs such as vehicle-emissions inspection and maintenance and the changes that actually occur. The evaluations should involve field observations (e.g., ambient air measurements, tunnel studies, and remote sensing), air-quality modeling, and vehicle-emissions data (e.g., data from vehicle-emissions inspection and maintenance programs, roadside pullover inspections, and other direct on-board tailpipe emissions measurements). Emphasis should be placed on techniques that are considered to have the fewest and smallest uncertainties.

Sensitivity and Uncertainty Assessment

Finding

At present, the understanding and quantification of the uncertainties in MOBILE are inadequate. There are uncertainties in the data used to develop model algorithms, the statistical analysis of test data, and the model input parameters. All of these lead to large uncertainties in model outputs.

Further, the committee is unaware of any specification of the level of accuracy required from MOBILE to support specific decision-making processes, in contrast to the model performance guidelines for travel-demand and air-quality modeling. The level of required accuracy would be expected to differ depending upon the various types of decisions that are required. Uncertainty and sensitivity analyses of MOBILE should focus on improving those elements that have the most impact on model results to help guide the development of testing programs and the next generation of models.

Recommendation

EPA, along with other agencies and industries, should conduct sensitivity and uncertainty analyses of the mobile-source emissions models in the toolkit, especially MOBILE, and explicitly assess the required accuracy for specific applications. These analyses should occur as a part of the ongoing process of model development and updating. Specifically, the analyses should:

- include a rigorous study of the model's sensitivity to all the input data to provide users with information on the most critical factors affecting model results;
- include a rigorous study of uncertainties and bias in all model components and in the data used to develop model parameters and relationships;
- explicitly define the levels of accuracy needed to fulfill EPA's regulatory responsibilities; and
- be used to design future versions of MOBILE and other models in the toolkit.

Long-Term Planning

Finding

EPA has not engaged adequately in long-term planning to coordinate future model-application needs with model developments. In general, the large transportation and emissions modeling efforts by EPA, the California Air Resources Board, the U.S. Department of Transportation, and others are not sufficiently integrated to make the most effective use of data, techniques, and other resources. The result of this lack of coordination and cooperation is that comparable models from different agencies are inconsistent.

Recommendation

EPA should promptly improve long-term planning to increase coordination in mobile-source emissions modeling. Within 1 year, EPA should coordinate with the California Air Resources Board, the U.S. Department of Transportation, and others to complete a long-range plan that addresses improvements of or new approaches to mobile-source emissions models. EPA should also arrange for a short policy study by an independent body to identify possible institutional barriers and ways to enhance institutional coordination in model development.

EPA should develop partnerships with other public and private organizations, in the United States and internationally, to improve planning and coordination of model development. This planning process should include the following:

- analysis of policy and modeling needs relevant to the control of ozone, hazardous air pollutants, and particulate matter (PM) and the development of off-road emissions regulations;
- analysis of modeling techniques that might be applicable to future versions of MOBILE and other mobile-source emissions models;
- assessment of present and future data needs and the feasibility of coordination of data-collection efforts;
- strategies and methods to improve the linkages among transportation, mobile-source emissions, air-quality, and exposure models;
- better use of advances in supporting technologies (e.g., Geographical Information System (GIS) and other information systems); and
- development of methods for assessing model accuracy.

Improving Characterization of Real-World Vehicle Emissions

Finding

A critical element in model accuracy is the accurate measurement of the emissions of current and new vehicles that are in use. Such measurements are especially important for developing emissions inventories for future years. EPA has projected substantial reductions in future deterioration rates of emissions-control equipment on the current fleet and has applied this trend to the emerging new generation of vehicles. Vehicles with accelerated mileage accumulation have been relied upon to some extent for predicting lifetime performance of technologies, although there is no assurance that vehicles with accelerated mileage accumulation accurately represent on-road vehicles deteriorating through normal aging and mileage accumulation.

Emerging technologies, such as on-board diagnostic systems that detect emissions-control-system failures, offer the possibility of significant future emissions reductions. The committee is unaware of any studies conducted to assess the motorist response to illuminated malfunction indicator lights, although this response rate is one of the most important factors for determining the effectiveness of on-board diagnostic systems in reducing emissions. It is also important to understand whether deterioration rates for emerging emissions-control technologies will be similar to those of in-use technologies or whether there are inherent differences that will result in vehicle emissions deteriorating at faster or slower rates.

Recommendation

EPA should develop a program to enable more accurate determination of in-use emissions. This program should use more real-world approaches such as direct tailpipe-emissions monitoring systems and random roadside pullovers of vehicles (such as those done in California) to ensure accurate characterization of emissions. Estimation of deterioration rates should be based on age as well as mileage. This might be difficult to do because of the correlation between age and mileage in the vehicle data used for MOBILE.

The program should include the development of improved estimates for various parameters in the MOBILE model. The parameters in the inspection and maintenance portions of the model, such as repair effectiveness, mechanic training, and deterioration of repaired vehicles are particularly important. A critical parameter in estimating the effectiveness of future inspection and maintenance programs is the fraction of motorists that get repairs in response to malfunction indicator lights of on-board diagnostic systems. The response rates must be established for areas with and without vehicle-emissions inspection and maintenance programs. This evaluation could be aided by use of a permanent on-board diagnostic memory system and simple data download techniques.

Appropriate statistical approaches should be used to project how vehicle emissions will change as vehicles normally age. Further, techniques should be developed to better capture the effects of failures of future control systems. This information will allow improved forecasting of the emissions for future model years with new emissions-control technologies.

RECOMMENDATIONS FOR IMPROVEMENTS TO MOBILE

The committee has developed a set of recommendations for the improvement of MOBILE as an air-quality planning tool. The most significant improvements that can be made to MOBILE are the acquisition of data necessary to improve the accuracy of the model and validation and evaluation of the model.

Emissions from Heavy-Duty Diesel Vehicles

Finding

NO_x emissions from heavy-duty diesel vehicles are underestimated in the current MOBILE model, and both NO_x and particulate matter (PM) emissions rates are highly uncertain. The proposed MOBILE6 emissions factors use engine certification data (in grams per brake horsepower hour) and conversion factors to estimate gram per mile emissions factors. For MOBILE6 to improve its accuracy, the model must better characterize real-world emissions from this vehicle class. The model should be upgraded soon, because State Implementation Plans that depend on correct assessments of NO_x control levels are now being developed and submitted. States are also developing plans to address problems with fine (less than 2.5 mm diameter) PM emissions, and emissions from heavy-duty diesel vehicles are expected to be a major target of those plans.

Recommendation

EPA should design and undertake a large-scale testing program that will better assess real-world emissions from heavy-duty diesel vehicles. The results should be incorporated into a subsequent revision of MOBILE6. This testing program should include a broad range of engine technologies and ages and be based on driving cycles that accurately reflect real-world driving patterns.

Particulate Emissions

Finding

PART5 is inadequate for supporting the new PM ambient air-quality standards and regional haze regulations. Although the results of field studies are conflicting, they indicate that PART5 does not provide an accurate current inventory of emissions. Other concerns with PART5 are its estimation of the effects of emissions-control equipment deterioration on PM emissions from heavy-duty diesel vehicles and emissions of fine PM. Given that EPA and the California Air Resources Board need to improve their modeling of PM emissions, early cooperation between the two agencies would produce a more unified approach and a stronger database.

Recommendation

EPA should promptly update PART5 with the best available data on PM emissions and incorporate it into a subsequent revision of MOBILE6. PART5 should be substantially upgraded and evaluated against field studies. These improvements should be carried out in collaboration with the larger model-development community. In particular, improvements in modeling PM emissions should be coordinated with the California Air Resources Board's efforts in modeling these emissions. Testing data are available that would greatly improve PART5 model predictions. However, EPA, with input from the technical community, should still assess data gaps and design and implement test programs to fill those data gaps. Of special concern should be the modeling of emissions of fine PM.

High-Emitting Vehicles

Finding

Emissions from high-emitting vehicles are a large source of uncertainty, and EPA has been slow to characterize these high-emitting vehicles. There is a lack of data on the emissions, number, and activity of these vehicles as well as a lack of information characterizing the effects of vehicle age, model, and geographical region. These vehicles are thought to represent a substantial fraction of mobile-source emissions and are the focus of some emissions-control programs, so they must be accurately characterized. Random roadside pullover testing of exhaust emissions, such as those currently performed in California, appears to be one of the most promising means of identifying vehicles with high exhaust emissions. Remote sensing of exhaust emissions has shown some promise as well.

However, neither of those techniques can be used to estimate emissions from vehicles with unusually high levels of evaporative emissions. VOCs can evaporate from a vehicle fuel system or result from liquid leaks. Improved characterization of emissions from vehicles with high evaporative losses might be critical for resolving differences between MOBILE-estimated and observed emissions of VOCs.

Recommendation

EPA should begin a substantial research effort to characterize high exhaust and evaporative emitting vehicles. These vehicles should be characterized in terms of number of vehicles, proportion of the on-road fleet, emissions rates, and travel-activity patterns. Effects of emissions-control programs, especially vehicle emissions inspection and maintenance programs, on high-emitting vehicles must also be properly assessed and modeled. MOBILE currently assumes a uniform distribution of vehicles in a region. However, high-emitting vehicles represent a disproportionate fraction of the total vehicles in low-income areas. To model the effects of differences in spatial and temporal distribution of high-emitting vehicles, local planning agencies need to gather such distribution data and conduct separate runs of MOBILE for various subareas in their planning region.

Frequency of Model Updates

Finding

Updating MOBILE5 to MOBILE6 has taken far too long. Information that invalidated assumptions in MOBILE5 about the deterioration of light-duty vehicles and the effectiveness of vehicle emissions inspection and maintenance programs, oxygenated fuel, and other control programs has been available for several years and has not been incorporated into the model. Thus, emissions inventories and control strategies being developed are based on out-of-date assumptions and inaccurate predictions, perhaps resulting in the selection and propagation of inefficient or ineffective controls.

Recommendation

EPA should be more timely (perhaps 1 to 2 years) in updating significant individual components of the model as important new information becomes available. Consolidated documentation written for end-users that explains how MOBILE works, how the components were updated, and how the new data sources are used should accompany model updates.

Mobile-Source Toxic Emissions

Finding

Although EPA has developed a model for predicting mobile-source toxic emissions based on the MOBILE5 model, this model is not publicly available. When this model is released, it probably will be the best available tool for assessing both mobile-source toxic emissions and the impacts on toxic emissions from control programs designed to reduce primary mobile-source pollutants.

Recommendation

EPA should incorporate estimates of mobile-source toxic emissions into MOBILE. The best available data should be used to update MOBTOX, which should be merged into MOBILE6. EPA should assess weaknesses in the mobile-source toxics databases, design and run test programs to fill data gaps, and incorporate new test data into a timely update of MOBILE6.

OTHER RECOMMENDATIONS

Off-Road Emissions

Finding

As emissions from on-road vehicles decrease due to tighter emissions standards, fuel-sulfur controls, and less deterioration of emissions-control devices, the emissions from off-road mobile sources will continue to increase in importance. That is particularly true for NO_x and PM emissions. Although the committee's charge did not explicitly call for an evaluation of EPA's new model for off-road-emissions (NONROAD), the committee recognizes the importance of accurately predicting off-road-emissions for evaluation of human health and environmental impacts from mobile-source emissions. Primarily because of a lack of data, the current off-road-emissions model does not accurately estimate off-road emissions inventories or the effects of emissions controls on these sources.

Recommendation

Within 1 year of the release of MOBILE6, EPA should have a plan for compiling the needed data and using these data to update NONROAD.

The plan should include the population and activity data and real-world emissions factors for gasoline and diesel engines.

TAKING THE NEXT STEPS

The recommendations presented in this report will not be easy to accomplish. Coordination of data collection, modeling, and evaluation efforts by EPA, the California Air Resources Board, the U.S. Department of Transportation, and others will need to be increased substantially. Significant resources will be needed to improve mobile-source emissions modeling. For example, the accuracy of MOBILE is limited by the availability of appropriate emissions testing, and the task of improving the emissions database is statistically complex and requires extensive resources. Expenditures of resources for testing, modeling, and evaluation are warranted, however, because the decisions that rely on the results of this model are of tremendous importance to human health, the environment, and the economy. Failure to provide the required resources will likely result in a continued loss of confidence in the accuracy of MOBILE and possibly result in inappropriate allocation of resources for mobile-source emissions controls.

Recent reorganization of the EPA Office of Mobile Sources, the EPA office responsible for MOBILE, into the Office of Transportation and Air Quality might have an impact on MOBILE's continued development. The reorganization should not impede rigorous development of MOBILE, which must be seen as an accurate reflection of mobile-source emissions.

1

Overview of Mobile-Source Emissions

N ACCURATE ASSESSMENT of motor-vehicle emissions is essential for an effective air-quality improvement program. The U.S. Environmental Protection Agency's (EPA's) Mobile Source Emissions Factor (MOBILE) model is the primary tool used by air-quality planners at national, state, and local levels to estimate on-road mobile-source emissions, and hence is key to assessing associated environmental impacts. Because of the model's importance in assessing air-quality control programs and because of concerns about weaknesses in the accuracy and reliability of the model, Congress asked the National Academy of Sciences to review and evaluate the MOBILE model. The National Research Council's Committee to Review EPA's Mobile-Source Emissions Factor Model was formed in response to that request. Specifically, the committee was to consider the adequacy of the model's input data, assumptions, structure, and results and recommend ways to improve the reliability of its mobile-source emissions estimates.

This chapter reviews mobile-source emissions within the context of the total air-quality problem in the United States, provides an overview of the categories and estimated levels of mobile-source emissions, introduces the legislative and regulatory initiatives used to control such emissions, details the committee's charge, and describes the structure of this report.

AIR-QUALITY PROTECTION

Air-quality policy in the United States is directed largely towards protecting public health. It also addresses public-welfare issues, such as visibility in national parks and wilderness areas and environmental damage due to acid deposition. The National Ambient Air Quality Standards (NAAQS) sets a primary standard for ambient concentrations of criteria pollutants to protect public health with "an adequate margin of safety", and a secondary standard to protect public welfare against environmental and property damage. The attainment of NAAQS for the six criteria pollutants, ozone, carbon monoxide (CO), nitrogen dioxide (NO_2), lead, sulfur dioxide (SO_2), and particulate matter (PM), is one measure of air quality.

Another category of pollutants consists of the hazardous air pollutants or "air toxics," including carcinogens, neurotoxins, teratogens, allergens, and other harmful compounds. Control of air toxics is aimed at ensuring that any health risks are less than one in a million from lifetime exposure to a particular toxic.

Other air-quality protection programs are directed towards acid-rain and visibility concerns. Acid-deposition regulations are designed to reduce the negative impacts on humans and the environment from the deposition of nitric and sulfuric acids. Visibility regulations are designed to protect the visibility in federal Class I areas, which are primarily national parks and wilderness areas.

Pollutants of Interest

The pollutants of interest for air quality are both primary and secondary pollutants. Primary pollutants are those directly emitted to the atmosphere and include CO, SO_2, and lead. Ambient concentrations of such pollutants are directly related to their sources. Secondary pollutants are those formed by atmospheric processes, including chemical reactions and condensation. Ozone is a secondary pollutant, formed by the action of sunlight and chemical reactions involving volatile organic compounds (VOCs)[1]

[1]An organic compound is a compound containing carbon combined with atoms of other elements, commonly hydrogen, oxygen, and nitrogen. Simple carbon-containing compounds such as carbon monoxide and carbon dioxide are usually classified as inorganic compounds. A volatile organic compound is a compound that can exist as a gas under typical atmospheric conditions. This report, unless otherwise noted, will refer to the general class of gaseous organic compounds, as VOCs. Appendix B describes the differences among the terms used to refer to gaseous organic compounds.

and nitrogen oxides (NO_x). Airborne PM and air toxics are combinations of primary and secondary pollutants.

In urban areas, motor vehicles generally are the dominant emissions sources of VOCs, NO_x, and CO and their control is critical for reducing urban air-pollution problems caused by these emissions. The adverse effects of ozone and CO are well understood, whereas those of PM and most toxics are less well understood. As more is known about the specific aspects of the toxicity of PM and air toxics, a better understanding of the contribution of motor vehicles to these pollutants becomes important.

Mobile-Source Contributions

Total mobile source emissions (on-road plus off-road emissions) contribute significantly to overall air pollution in the United States.[2] Figure 1–1 shows estimates for 1997[3] of the contribution of mobile sources (on-road and off-road) to emissions of criteria air pollutants and their precursors. For example, mobile sources in that year contributed over 75% of CO emissions (about 67,000 thousand tons), almost half of the NO_x (about 11,600 thousand tons), and 40% of the VOCs (about 7,700 thousand tons) (see Figure 1–1) (EPA 1998b). The elimination of lead from gasoline has greatly reduced mobile-source emissions of this pollutant to just over 500 tons (13% of total lead emissions) in 1997, compared with mobile-source lead emissions in 1970, which were over 180,000 tons (EPA 1998a). According to the EPA (1998b), mobile-source exhaust is a less important source of PM-10 (those particles smaller than 10 mm in diameter) and sulfur dioxide. Mobile-source contribution to fine particles (those smaller than 2.5 mm in diameter and referred to as PM-2.5) is an area of continuing study. One recent study reported higher than expected PM-2.5 emissions from light-duty vehicles (LDVs) at higher elevations (Cadle et al. 1998). It should be noted that, for a given location, the fraction of emissions inventories contributed by mobile sources varies greatly.

---

[2]The principal sources of emissions are stationary sources (fuel combustion by utilities, industry and residential sources), industrial process sources (chemical manufacturing, petroleum refining, solvents and waste disposal), on-road vehicles (light- and heavy-duty gas and diesel-powered vehicles), non-road engines and vehicles (recreational, industrial, and commercial vehicles and engines, trains, marine vessels, and aircraft), and other (biogenic emissions from natural, agricultural and forestry sources, and other combustion) (EPA 1998a).

[3]To the extent that the mobile source emissions estimates are obtained by using the MOBILE model, they are subject to the types of uncertainties discussed throughout this report.

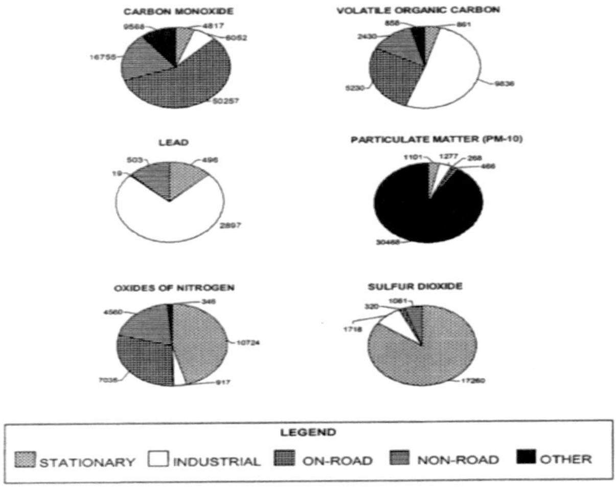

FIGURE 1–1 Sources of Criteria Air Pollutants. Estimated total annual emissions of criteria pollutants from stationary, industrial process, and mobile (on-road and non-road), and other sources for 1997. Emissions are shown in thousands of tons except for lead, which is shown in tons. Source: EPA 1998b.

Human Health Concerns

Total emissions from mobile sources contribute significantly to the detrimental health effects resulting from exposure to ambient ozone, CO, PM, and air toxics. Urban ozone has been one of the most persistent health concerns. The current 1-hr primary NAAQS for ozone is 0.12 parts per million (ppm), which is the daily maximum not to be exceeded more than once per year on average. (This is averaged over 3 years, so the fourth highest 1-hr value over 3 years is the one that is used to compare with the standard.) Health effects associated with exposures above this standard are well documented and are summarized by Lippmann (1989). They range from short-term consequences such as chest pain, decreased lung function, and increased susceptibility to respiratory infection, to possible long-term consequences, such as premature lung aging and chronic respiratory illnesses. In 1997, approximately 48 million U.S. residents in 77 counties, primarily urban and suburban regions, lived in areas where the second highest daily maximum concentration exceeded 0.12 ppm (EPA 1998b). Due to concerns about prolonged exposures to lower levels of ozone, EPA adopted an 8-hr standard of 0.08 ppm, which would put a much greater area and population into nonattainment with the standard (Wolff 1996; Chameides et al. 1997). Based on the period from 1993 to 1995, EPA estimated that 248 counties with a population of 83 million people would have violated the 8-hr ozone standard (EPA 1997a). Prospects for the implementation of the new standards are uncertain because a U.S. Court of Appeals in May 1999 remanded them for further consideration by EPA.

Attaining CO standards has been far more successful. There are two primary standards for CO, a 1-hr average of 35 ppm and an 8-hr average of 9 ppm. The health effects from exposures to concentrations exceeding these standards are also documented. CO enters the blood stream and links to hemoglobin, reducing the amount of oxygen the blood can carry and causing mental and physical impairment. In 1997, approximately 9 million people in three urban counties lived in areas that exceeded these standards (EPA 1998b). This number has been declining over time.

Standards for PM and air toxics are currently changing due to increased knowledge about their health effects. Current regulations for ambient concentrations of PM focus on particles less than or equal to 10 mm. New regulations to control smaller particles (less than or equal to 2.5 mm), intended to address some preliminary findings that these particles have the greatest impact on human health (Dockery et al. 1993), were remanded for consideration back to EPA by a U.S. Court of Appeals in May 1999. Air toxics include a wide class of emissions and effects. Few of these substances have been studied to examine possible health effects. In 1998, the California Air Resources Board (CARB) identified diesel PM as a toxic air contaminant.

An issue unique to mobile sources is the proximity of sources to receptors. Air vents on cars can scoop up exhaust from a vehicle just ahead so that exposures in the cab of a car or truck are higher than at the roadside. A jogger at a roadside, breathing deeply and rapidly, might inhale much larger amounts of contaminants than a sedentary individual sitting out-doors away from the road. Urban canyons can retain, and thus concen-trate, contaminants emitted by motor vehicles. Figure 1–2 shows how exposure to CO varied for one individual during typical commutes. Because health effects are a function of what is actually inhaled, this variation in exposure is a critical feature of mobile-source pollutants.

Environmental Concerns

Mobile sources contribute significantly to the risk of detrimental environmental effects of ozone, acidic deposition, PM, and air toxics. However, environmental impacts have received much less attention than health effects. For most criteria pollutants, the secondary NAAQS, set to protect the environment and susceptible outdoor structures, is the same as the primary standard set to protect human health. Ozone concentrations can damage crops and other vegetation (Heck et al. 1982; EPA 1986), exhaust emissions are toxic to roadside flora and fauna (particularly evident when lead was used (EPA 1998b), acid can damage human artifacts (e.g., stain-ing of buildings and corrosion of outdoor statuary) (NAPAP 1990), and particulates contribute to haze and poor visibility (NRC 1993).

Mobile sources also contribute to greenhouse gas emissions. Approximately one-third of the total U.S. anthropogenic emissions of CO_2 comes from the transportation sector (Energy Information Administration 1997), including about one-quarter of the total from light-duty vehicles and heavy-duty vehicles (Heinz Center 1998). Also, mobile-source emissions of chlorofluorocarbons and hydrochlorofluorocarbons from air-conditioning systems contribute to stratospheric ozone depletion. Although chlorofluorocarbons are now banned from use in the United States, existing systems containing these compounds will continue to leak them to the atmosphere (Holmes and Ellis 1997), and significant illegal use has been reported (Valletta 1995).

ESTIMATING EMISSIONS FROM MOBILE SOURCES

Importance of Source Identification and Quantification

An effective air-quality improvement program requires the identification, inventory, and control of emissions sources, including mobile sources.

This requires not only a broad understanding of which pollutants are derived from which sources, but also details about their spatial and temporal variation, the contributions of subsets of sources, the chemical and physical characteristics that determine their propensity to form secondary pollutants, their levels of exposure and toxicity, and the actual effectiveness of strategies to control emissions. The large number of individual sources, the large variability of emissions characteristics among these sources, and the need for emissions estimation methods to fulfill many applications creates daunting challenges. To aid in identifying, estimating, and reducing risks of motor-vehicle emissions, EPA has developed a series of models referred to as the MOBILE model. The current (MOBILE5b) and upcoming (MOBILE6) versions of MOBILE provide some but not all of the required information. MOBILE estimates emissions factors from broad vehicle classes using average vehicle speed to represent highway conditions. However, MOBILE provides only limited information on the physical and chemical characteristics of emissions, and it does not simulate emissions related to dynamic traffic-flow conditions.

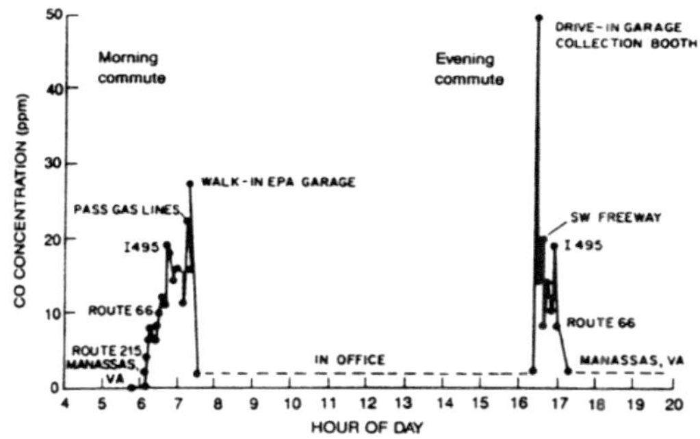

FIGURE 1–2 Daytime exposure to carbon monoxide during peak-hour commutes. Source: Ott 1985.

The MOBILE model deals only with the on-road vehicle emissions of CO, VOCs, and NO_x. Related models cover other mobile-source emissions (e.g., NONROAD for off-road emissions, PART5 for PM, and MOBTOX for air toxics). These related models are often discussed in the same context as MOBILE because they are needed to represent the full suite of emissions from mobile sources and often use the same input data as MOBILE.

On-Road Vehicle Emissions Categories

Regulations and the MOBILE model divide on-road vehicles into two broad categories, light-duty (LDVs) and heavy-duty (HDVs) vehicles. The size boundary between LDVs and HDVs historically has been 8,500 pounds, gross vehicle weight (the weight of the vehicle plus the weight of the rated load-hauling capacity). LDVs are fueled primarily by gasoline. HDVs are fueled by diesel and gasoline with the heavy HDVs (those with gross vehicle weights greater than 26,000 pounds) fueled with diesel. These heavy HDVs are important because, though they represent a small fraction of the total number of vehicles, they represent a significant fraction of total vehicle miles traveled, fuel consumption, and emissions (Davis 1999; EPA 1998a). Figure 1–3 displays the separation of emissions by fuel use.

Light-Duty Vehicles

Historically, LDVs were sub-divided into two main categories[4]: passenger cars and light-duty trucks (LDTs), with the latter used primarily for commercial purposes. Two different sets of exhaust emissions standards applied. Differences between the two categories of vehicles have diminished and the regulatory distinction will disappear in 2007. Tailpipe emissions standards were applied in 1968 and now call for a reduction of more than 90% from 1968 levels for all pollutants.

Passenger cars, as the name implies, refer to personal vehicles used primarily to transport people. A single set of emissions standards applies to all passenger cars, regardless of size, passenger occupancy, or use. Emissions are regulated on the basis of grams of pollutant per mile (g/mi) and vehicles are certified on a chassis dynamometer test.[5] Table 1–1 shows historical categories of standards for passenger cars for up to 50,000 miles or from 50,001 to 100,000 miles.[6]

---

[4]Motor cycles also form a category of on-road vehicles. However, their emissions tend to be small relative to passenger cars and light-duty trucks. Thus, they are not discussed here.

[5]In a chassis dynamometer test, the whole vehicle is mounted on a dynamometer for testing. In contrast, in an engine dynamometer test only the engine, rather than the whole vehicle, is mounted on a dynamometer.

[6]Manufacturers are allowed to certify compliance using low mileage cars and an agreed-upon deterioration assumption. They are not required to recall and test inservice vehicles. (Recalls may be required if emissions control systems are shown to be faulty.)

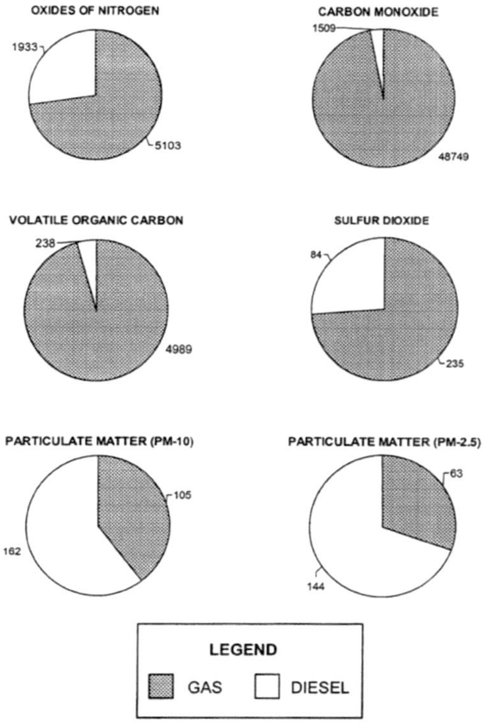

FIGURE 1–3 Estimated mobile-source emissions by fuel type. MOBILE5 and PART5 estimates of 1997 emissions from the on-road motor-vehicle fleet. It is likely that MOBILE5 underestimates gasoline VOC and diesel NO_x emissions. Emissions are shown in thousands of tons. Source: EPA 1998a.

TABLE 1–1 Passenger-Car Exhaust Gaseous Emissions Standards (g/mi)

	50,000 miles			100,000 miles		
	THC (NMHC)[a]	CO	NO_x	NMHC[a]	CO	NO_x
Model Year						
Pre-control	10.6	84.0	4.1			
1968–71	4.1	34.0	3/4	3/4	3/4	3/4
1972–74	3.0	28.0	3.1	3/4	3/4	3/4
1975–76	1.5	15.0	3.1	3/4	3/4	3/4
1977–79	1.5	15.0	2.0	3/4	3/4	3/4
1980	0.41	7.0	2.0	3/4	3/4	3/4
Category						
Tier 0 (1981–93)	0.41	3.4	1.0	3/4	3/4	3/4
Tier 1 (beginning with model year 1994–)	0.41 (0.25)	3.4	0.4	0.31	4.2	0.6
NLEV (beginning with model year 1999–)	3/4	3/4	3/4	0.09	4.2	0.3
Tier 2—Default set in CAAA90 (beginning with model year 2004–)	3/4	3/4	3/4	0.125	1.7	0.2
Tier 2—Current proposed standards (beginning with model year 2004–)	3/4	3/4	3/4	>0.09[b]	>4.2[b]	0.07

[a]Note: Emissions standards were originally written for total hydrocarbons (THC) and later for nonmehtane hydrocarbons (NMHCs). Appendix B describes the differences among the terms used to refer to gaseous organic compounds. This report, unless otherwise noted, will refer to the general class of gaseous organic compounds as VOCs.

[b]Note: The proposed Tier 2 standards are a corporate average standard with a focus on NO_x emissions. This allows NMHCs and CO emissions standards to "float", in that fleet emissions rates depend on the mix of vehicles used to meet the NO_x standard. The emissions standards shown for NMHCs and CO are those that would result given the mix assumed in the Notice of Final Rulemaking (EPA 1999a) to meet the NO_x standard.

Sources: EPA 1998c,d, 1999a; Davis 1997; Chrysler Corp. 1998.

The LDT category originally described vehicles designed for load-hauling rather than for passenger transportation, and was further divided into weight categories. Some of these LDTs, and some passenger cars, have evolved into vans and sport-utility vehicles. As a result, differences in size and function among passenger cars, LDTs, vans, and sport-utility vehicles have diminished. As a consequence, EPA in its Final Rule for Tier 2 emissions standards (EPA 1999a) mandates the same emissions standards to passenger cars and LDTs.[7] As with passenger cars, pollutant emissions limits for LDTs are expressed in grams of pollutant per mile and vehicles are certified on a chassis dynamometer. The MOBILE model retains the historical distinction between passenger cars and LDTs because current emissions standards and control technologies differ among vehicle classes.

Heavy-Duty Vehicles

Different exhaust emissions standards apply to the two broad categories of HDVs: trucks and buses, with further differentiation made between gasoline and diesel engines. Emissions are regulated on grams of pollutant per brake-horsepower-hour (BHP-hr) because of the difficulty of devising reasonable grams per mile limits for the broad range of vehicles covered and the difficulty in developing a practical chassis dynamometer test. Engines are certified on an engine dynamometer. Emissions standards were applied later than for LDVs and are less stringent than for LDVs.

Because certification is for engines rather than vehicles, measurements are of engine emissions that are then used to estimate vehicle emissions. To obtain emissions in grams of pollutant per mile, which is what MOBILE calculates, vehicle efficiency (BHP-hr/mile) must be estimated. The problem of converting engine dynamometer emissions to on-road emissions is the same for both buses and trucks. Engines used in buses must meet more stringent PM standards than engines used in trucks.

Evaporative Emissions

Evaporative emissions, including those resulting from leaks of liquid fuel, can be classified into five categories for modeling purposes: diurnal,

[7]EPA has also introduced a new category of vehicle in the Tier 2 proposal, the medium-duty passenger vehicle for passenger vans and sport-utility trucks between 8,500 and 10,000 pounds gross vehicle weight. This category of vehicle will also be required to meet the Tier 2 emission standard.

hot soak, running loss, resting loss, and refueling loss. Originally, evaporative emissions were regulated for a combined sum of hot soak and diurnal emissions. In more recent years, a separate limit was placed on running loss evaporative emissions. The first on-board refueling-loss standard began with a 3-year phase-in period on passenger cars in 1998. An onboard refueling vapor canister controls these emissions. Another method of reducing refueling emissions, Stage II refueling controls, alters the design of gasoline service station nozzles to allow the collection of vapors displaced during refueling. Large uncertainties are associated with the ability to accurately characterize evaporative emissions. Of particular concern are large liquid leakers, which have been shown to result in extremely high VOC emissions for such vehicles (Haskew et al. 1999).

Off-Road Emissions Categories

Non-road, or off-road, vehicles have been less regulated, and have come to contribute a greater fraction of pollutants as emissions from on-road vehicles have been controlled. These emissions are estimated separately from the MOBILE model. Their contribution is becoming increasingly important in estimating total (on-road plus off-road) mobile-source emissions. Figure 1–4 shows an increasing percentage of NO_x emissions coming from off-road sources between 1970 and 1997. The wide variety of sources in this category include:

- ***Construction, logging, mining, and farm equipment***—This equipment uses engines similar to those in heavy-duty trucks, primarily diesel, but operating characteristics can vary.
- **Lawn and garden equipment**—This category includes small devices such as lawn mowers, chain saws, and leaf blowers. Both two-stroke and four-stroke gasoline engines are used.
- ***Recreational vehicles***—This category includes land and watercraft that use diesel and both two- and four-stroke gasoline engines.
- ***Industrial, light commercial, and airport services***—These vehicles are similar to those used for on-road and specialty applications, and use both gasoline and diesel engines in a variety of applications.
- ***Locomotive, marine, and aircraft engines***—Very large diesel and gas turbine engines dominate these applications. Emissions sources that do not impact air pollution are usually excluded, such as emissions from offshore ships and aircraft operations at high altitudes. Marine emissions associated with offshore oil operations are classified as stationary source emissions.

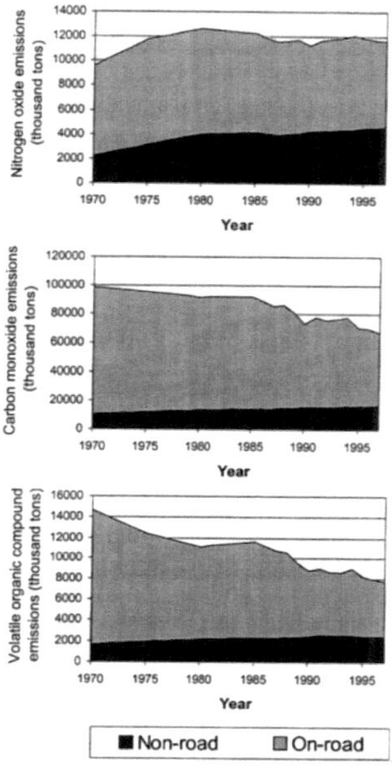

FIGURE 1–4 On-road and non-road emissions for 1970–1997. Linear interpolation of data between 1970–1975, 1975–1980, 1980–1985, and 1985–1987. Source: EPA 1998a.

Mobile-Source Emissions Using MOBILE and Related Models

MOBILE is but one of many models used for the estimation of motor-vehicle emissions. Details and limitations of MOBILE are discussed in Chapter 3, and alternative approaches are discussed in Chapter 5. The MOBILE model is used to estimate vehicle emissions factors that are combined with estimates of vehicle mix and activities to estimate emissions inventories. The emissions factors can be tailored to specific fuel compositions, vehicle technology types (i.e., noncatalyst, carbureted three-way catalyst, and port fuel injection), and operating modes (i.e., cold start and hot stabilized). MOBILE also adjusts emissions factors for parameters such as ambient temperature, altitude, average speed, and trip characteristics (i.e., numbers of cold starts and length of diurnal soaks). This information is developed for a default or specified fleet composition. The end result is an estimate of the average emissions per mile for a vehicle type and the vehicle fleet. Vehicle activity is specified in terms of miles traveled by vehicle type within some defined area. This information is combined with MOBILE's estimated emissions rate per mile to produce an emissions inventory for the area. Other related models use this information to estimate particulate and air-toxics emissions (PART5 and MOBTOX, respectively), the impacts of fuel composition (COMPLEX), and hydrocarbon speciation (SPECIES) profiles for air-quality modeling. Non-road mobile-source emissions are estimated in the NONROAD model by combining activity estimates for these sources with aggregate emissions factors. This modeling approach is much simpler than the approach used in MOBILE.

This report discusses several areas of concern about the use of MOBILE. The MOBILE model is now used for purposes for which it was not originally intended, such as providing detailed air-quality modeling inputs, selecting control strategies based on prospective emissions reductions, and demonstrating conformity of transportation projects with the Clean Air Act. The General Accounting Office (GAO) has pointed to problems arising from use of the MOBILE model by multiple users for multiple purposes (GAO 1997). Many of these uses are dictated by legislative and regulatory initiatives, discussed in the next section. Limited validation and the failure of the model to address uncertainty create additional problems. Some observed shortcomings of the MOBILE model are summarized in Table 1– 2, and discussed at greather length later.

LEGISLATIVE AND REGULATORY INITIATIVES

Legislative Requirements and Compliance Attainment Plans

The Clean Air Act and its amendments require that areas that have not met the NAAQS for ozone, carbon monoxide, nitrogen dioxide, sulfur diox ide, and inhalable particulate material develop plans, known as State Implementation Plans or SIPs, describing how they will attain compliance. The widespread ozone and PM problems in urban areas have driven much of the clean air planning and regulation nationwide. The Clean Air Act as amended in 1990 (CAAA90) requires a comprehensive attainment SIP from every ozone nonattainment area classified as serious, severe, or extreme. The Act prescribes certain minimum control measures for each ozone nonattainment area, based on the severity of the problem. The Act also prescribes technical criteria; for example, each plan must contain a current emissions inventory, adequate ambient air-quality data, and an analysis of future air quality based on photochemical grid modeling. Outside of California, the MOBILE model is used to estimate emissions from on-road mobile sources as part of the SIP. To ensure a minimum rate of progress, each ozone SIP must specify emissions targets for identified milestone years. The milestone years which emissions targets must be established include the attainment year and every third year of progress toward attainment.

TABLE 1–2 Issues Relating to the Use of MOBILE Discussed in This Report

Issue	Chapter
• Multiple uses and users need more than MOBILE can provide. EPA, DOT, state and local agencies, and the automobile and oil industries have differing needs.	Chapter 2
• Higher resolution of spatial and temporal scales are needed to evaluate the effectiveness of control strategies; that is, to provide estimates by time of day, day of week, and sublocations.	Chapter 2
• Models usually estimate emissions for only a few pollutants, specifically CO, VOCs, NO_x, PM (from the PART model), and air toxics (from the MOBTOX model). The impacts of these emissions vary with the chemical composition of the species and the size of the particles.	Chapter 3
• Modeled effectiveness of emissions control programs, such as motor vehicle inspection and maintenance, reformulated gasoline, and oxygenated fuels, are not directly linked to the measured effectiveness of such programs.	Chapters 3 and 4
• Estimated emissions do not correspond to field observations. Observations of the in-use fleet include remote sensing and tunnel measurements.	Chapter 4

Legislative Requirements—Conformity Plans

The CAAA90 prohibits regionally significant transportation activities, regardless of funding source, from impeding a region's progress toward attainment of the NAAQS. The conformity requirement is intended to ensure that emissions associated with transportation improvements are completely accounted for in the SIP. It is also to ensure that these improvements will not cause or increase the frequency of air-quality violations or delay attainment. The U.S. Department of Transportation (DOT) and metropolitan planning organizations (MPOs) determine whether projects pass the Conformity Demonstration. The procedures for Conformity Demonstration have been amended several times since the final conformity rule was issued by EPA (EPA 1993a).

As with attainment plans, the MOBILE model is matched with travel activity (e.g., trips, vehicle miles traveled, and speed profiles) to quantify the emissions of either a regionally significant project or a regional transportation plan.

Regulatory Initiatives

EPA has the authority to adopt standards for a broad range of mobile sources to assist states in achieving the NAAQS. For example, EPA used the CAAA90 to set tailpipe emissions standards for cars and light trucks beginning with the 1994 model year. These amendments require EPA to determine whether further emissions reductions from these vehicles are necessary. In 1998, EPA concluded that more stringent vehicle standards—known as Tier 2 standards—are needed to meet the NAAQS for ozone, and that the technology to meet these vehicle emissions standards is available and cost-effective. Along with the Tier 2 vehicle standards, EPA is lowering the allowable levels of sulfur in gasoline (EPA 1999a). Automobile manufacturers successfully argued that large reductions in gasoline sulfur content were needed to enable emissions control equipment to attain the Tier 2 emissions standards. EPA has also set emissions standards for medium- and heavy-duty gasoline vehicles, and is considering increasing the stringency of some of these standards in the near future. MOBILE is the tool used to estimate the emissions benefits from all of these regulatory initiatives.

EPA has also focused efforts on controlling emissions from heavy-duty diesel trucks and buses. Control of these emissions is important because diesel-fueled engines emit a complex mixture of gases, vapors, and particles. These include NO_x, PM, and many toxic air contaminants, including benzene, aldehydes, nickel, and polycyclic aromatic hydrocarbons. The fine particles produced by diesel engines are small enough to be inhaled and lodge deep in the lungs. Lower emissions standards for new heavy-duty diesel engines will be effective for the 2004 model year (EPA 1997b), and EPA is considering more stringent requirements for sulfur levels in diesel fuel and more stringent NO_x and PM standards for future diesel engines. MOBILE and its related models play a major role in assessing these standards.

COMMITTEE'S CHARGE AND HOW IT ORIGINATED

EPA first issued general guidance on model review in 1989 from its Science Advisory Board. The board recommended that the MOBILE model's predictive capability could be enhanced through: (1) obtaining external stakeholder input; (2) documenting the model's explicit and implicit assumptions; (3) performing sensitivity analyses; (4) testing model predictions against laboratory and field data; and (5) conducting peer reviews. As noted by the GAO (1996), the EPA's policy on peer review had not been followed during the updating of previous and the current versions of MOBILE. However, EPA's Office of Transportation and Air Quality (formerly the EPA's Office of Mobile Sources) has made the review process an integral part of the development of MOBILE6. An emissions modeling work group was established as part of the Mobile Source Technical Review Subcommittee to offer technical advice on MOBILE6 development. In addition, the Office of Transportation and Air Quality has, outside the subcommittee process, developed a formalized procedure to obtain and respond to extensive stakeholder comments on MOBILE6 development.

Congress has taken an interest in the accuracy and reliability of MOBILE because of its implications for the design of vehicle inspection and maintenance (I/M) programs and its role in regulatory decision-making. These issues were at the forefront during a 1995 hearing of the House Subcommittee on Oversight and Investigation on the effectiveness of vehicle I/M programs and how the MOBILE model credits these programs (U.S. Congress 1995). These concerns prompted Congress to request not only the NRC study reported here, but also a study by GAO that describes the major limitations of the current MOBILE model and EPA's progress towards improving the model. The NRC committee has used the results of the GAO study (1997) as a source of information for reviewing current and future versions of MOBILE. Congress requested the National Research Council to review the MOBILE model in its 1997 appropriation for EPA. The task statement reads as follows:

The committee will review the U.S. EPA's mobile source emission factor model (MOBILE). The committee will consider the adequacy of the model's input data, assumptions, structure, and results used to characterize mobile source emissions. To the extent possible, the committee will consider ways to improve the reliability of the MOBILE model as a tool for assisting in the development of emission control strategies to meet air-quality goals. Specifically, the committee will evaluate the current and planned versions of the MOBILE model, as appropriate, with respect to the following aspects:

- The types of mobile sources addressed, particularly the amount and quality of data from under-represented and unrepresented categories of sources (e.g., heavy-duty vehicles) and emissions of increasing interest (e.g. particles smaller than 2.5 mm).
- The strategies and methods for future data gathering in partnerships with other researchers, such as state agencies, industry associations or others with relevant data, in order to increase the amount and range of data in the most cost-effective manner.
- Alternative data sources and analytical techniques currently used for similar purposes by others (e.g. the California Air Resources Board in their EMFAC and other related mobile source emissions models) or the German government.
- The latest developments in related areas of modeling and how such advances might be incorporated in a new version of the model.
- The feasibility and requirements for the incorporation of modal modeling in MOBILE7 to reflect the effect on emissions of a variety of driving conditions and vehicle technologies.
- To the extent practical, the overall accuracy of the current version of the MOBILE model in predicting emissions to the atmosphere.

REPORT STRUCTURE

This report is the committee's response to its charge. The Executive Summary presents the committee's main recommendations. These recommendations encompass improvements to the MOBILE model as well as improvements to the overall process for estimating mobile-source emissions. Chapter 2 describes how the MOBILE model is used in air-quality planning and regulation. Chapter 3 briefly discusses the history and the technical aspects of the model and describes the major changes to the model for MOBILE6. Chapter 4 considers the accuracy and uncertainties of the MOBILE model. Chapter 5 describes alternative modeling approaches, some of which could be used to improve the modeling of mobile-source emissions. Chapter 6 incorporates the materials from the previous five chapters in a proposal for a new approach to modeling mobile-source emissions that relies on a toolkit of models that are better suited to the wide range of applications currently covered by MOBILE.

2

Current and Possible Future Uses of MOBILE in Air-Quality Management

N EMISSIONS-FACTOR MODEL is fundamental for assessing the nature and magnitude of on-road motor vehicle emissions and their impacts on ambient air quality. In the United States, excluding California, the MOBILE model has been the only model used in policy and regulatory settings to simulate actual emissions from automobiles over widely varying scales of resolution. (California uses the Motor Vehicle Emissions Inventory modeling suite for the assessment of vehicle emissions and their controls, as discussed in Chapter 5.) MOBILE is used in the development of national, regional, and urban emissions inventories; the simulation of regional air chemistry and microscale dispersion of pollutants; the assessment of the effectiveness of control strategies; the documentation of emissions reductions in State Implementation Plans (SIPs); the assessment of air-quality impacts of transportation projects, including the demonstration of conformity of transportation and air-quality plans; and the assessment of air-quality impacts of transportation-control measures and projects.

Traditionally, the management activities of air-quality regulatory agencies at the local, state, and federal level have used MOBILE to estimate vehicle emissions. Increasingly, transportation agencies, principally state departments of transportation and local metropolitan planning organizations (MPOs), have become more reliant on MOBILE in fulfilling their new obligations under the Clean Air Act Amendments of 1990 (CAAA90) and the Intermodal Surface Transportation Efficiency Act of 1991 (ISTEA).

These acts expand requirements for state transportation departments and MPOs to assess the air-quality effects of transportation plans and projects. Also, the automotive and oil industries, consultants, and academic organizations use MOBILE in a variety of ways related to air-quality regulation and, more broadly, to develop a better understanding of the dynamics of atmospheric pollutants.

FUTURE MOBILE-SOURCE EMISSIONS-MODELING ISSUES

Originally, MOBILE was developed to estimate overall emissions levels, trends over time, and the effectiveness of mobile-source emissions-control strategies. The model has undergone significant evolution since then, which is summarized in Chapter 3. Current uses of the model include developing emissions inventories and reductions in SIPs, demonstrating conformity of transportation and air-quality plans, and providing emissions estimates for dispersion and photochemical air-quality modeling. Thus, the original role of MOBILE has been expanded in ways that now require higher standards of accuracy that incorporate a greater degree of complexity. A good example of this evolution is demonstrated by MOBILE's current use in ozone attainment modeling, which requires precise spatial and temporal estimation of speciated precursor emissions to predict ambient ozone levels for a particular day and region. This application is more difficult and demands greater precision than MOBILE1's use for modeling of exhaust emissions as a function of age or mileage.

The need for vehicle emissions models will continue in the future and the demand for more accuracy and versatility will increase. SIP requirements to meet proposed new fine particulate matter (PM-2.5) and 8-hr ozone standards, regional visibility rules, and increased interest in air toxics from mobile sources will magnify concerns about MOBILE's applicability and accuracy. Although MOBILE6, the upcoming version of MOBILE, will address some of these concerns, it will likely fall short of the regulatory burden placed on its use.

For instance, MOBILE6 assumes lower rates of deterioration of vehicle emissions-control systems than earlier versions because of indications that newer technology is more durable. Yet, as discussed in Chapter 3, the extent to which the data supporting this position is fully representative is questionable. A desire for more accurate microscale modeling of localized transportation-control measures and more accurate photochemical modeling will add to the growing need for instantaneous or modal emissions modeling and better spatial disaggregation of emissions. Disaggregation of certain model components, such as the separation of start emissions from running emissions or the development of facility-specific speed cor rection factors, will provide better resolution, but the activity data required to implement these changes may not be available in all regions. The demand for a more accurate MOBILE model also raises concurrent demands for more-accurate transportation-activity factors such as average vehicle speeds, congestion levels, roadway classification, or miles traveled. Finally, it is conceivable that future users could require a real-time version of a mobile emission model to manage traffic flow in order to minimize emissions in critical air-quality areas. No single model is available to accurately address all of these possible uses.

MODELING AIR QUALITY: AN INTERDISCIPLINARY ENDEAVOR

Efforts to evaluate the air-quality impact of on-road motor vehicles are inherently interdisciplinary, and require the interaction of three different models and related areas of expertise: travel-demand models, emissions models, and air-quality models. Travel-demand models determine the amount of transportation activity occurring in a region based on an understanding of the daily activities of individuals and employers as well as the resources and transportation infrastructure available to households and individuals when making their activity and travel decisions (Harvey and Deakin 1993). This includes measures such as number of trips, time of day, length of trip, mode of transportation, route or location of trips, average speed of travel, and age of vehicle. The number of transit trips, automobile occupancy, and vehicle miles of travel (VMT) are common performance measures used to measure transportation activity.

The second component corresponds to mobile-source emissions rates. MOBILE estimates emissions rates based on vehicle type, average speed, ambient temperature, and other factors. The product of the transportation activity and the emissions rates from MOBILE results in emissions estimates for each modeled pollutant (carbon monoxide (CO), volatile organic compounds (VOCs), and nitrogen oxides (NO_x)). It is critical that estimates of transportation activity and emissions rates be in balance with respect to fidelity, accuracy, and precision to ensure the reasonableness of the emissions estimates. It is impossible to have the same for both (in some cases, the vehicle activity estimates will be more precise, or more accurate, or more refined, and in others the emissions rates will be so). However, transportation and air quality planners should understand the fidelity, accuracy, and precision for each component and take these into account in policy analysis (such as through uncertainty analysis and development of confidence bounds).

The third component of the modeling trilogy is the regional and microscale modeling of air quality. These models translate emissions invento ries into ambient pollutant concentrations that vary through space and time. Translating emissions to ambient concentrations can be done directly, for example, by using microscale carbon monoxide modeling. This method estimates concentrations in "hot spot" areas (critical intersections and sites with violations or possible violations of the NAAQS) by simulating the dispersion of the pollutant, using a variety of dispersion parameters, such as wind speed and direction. This is also done through urban-scale and regional-scale air-quality models that calculate ozone concentrations by simulating both atmospheric chemistry and meteorology. Again, attention to fidelity, accuracy, and precision is needed in each of these three types of models to ensure balance in their integration.

Travel-Demand Modeling

There are five traditional components of the sequential travel-demand model, namely demographic forecasting and the four-step travel-demand modeling process (Figure 2–1).

- **Demographic data**—The location of households and employment (categorized as basic, retail, and service) in small traffic survey zones within the urban region. This includes the forecasts of regional economic growth, land use patterns, and future demographic trends.
- **Trip generation**—The estimation of the number of trips by zone by time of day and type (both trips originating in a zone, termed trip production, and trips terminating in a zone, termed trip attraction).
- **Trip distribution**—The pairing of trip productions with trip attractions resulting in a full spatial pattern of travel by purpose and time of day.
- **Mode choice**—The determination of mode of travel, specifically walk, bicycle, drive alone, high occupancy vehicle (HOV), bus, rail, or truck travel.
- **Route assignment or choice**—Trips are assigned to paths in the transportation infrastructure by minimizing travel times or travel times and costs, and incorporating average speed and other impedance feedbacks.

The travel-demand models answer: "Will I travel," "How often and when," "Where," "By which mode," and "By which route"? They provide the MOBILE model with information on average vehicle speeds for each roadway segment that may be aggregated by roadway type or facility (e.g., freeways, arterials, collectors, and freeway ramps). The following points describe some important issues that relate to the use of travel-demand modeling in modeling air quality.

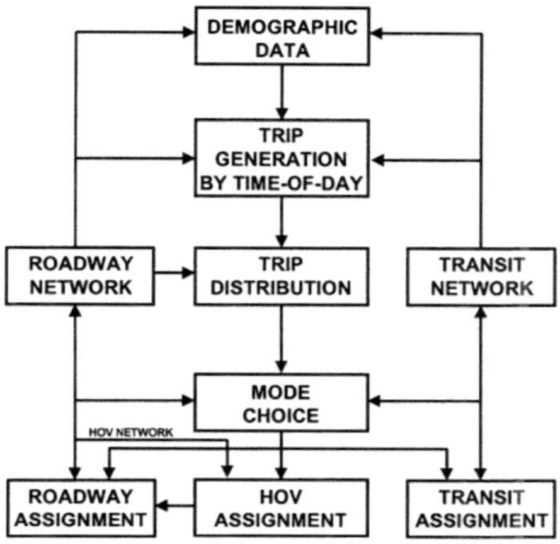

FIGURE 2–1 Sequential travel-demand forecasting process used in Dallas, Texas. Source: NCTCOG 1999.

- **Seasonal variations**—Most travel models are calibrated for typical weekdays when primary and secondary schools are in session. Therefore, travel is modeled for a typical weekday in February through May and September through November. However, air-quality assessments typically do not follow such convenient transportation schedules, because ozone is a frequent summer problem and carbon monoxide is often a winter one. Therefore, the travel-activity information may be modified to model the emissions type and season of interest.
- **Adjustments for weekend/weekday**—Travel models typically simulate weekday traffic, with adjustments for weekend emissions inventories often being required for air-quality assessments. Travel survey data is being collected to assist estimating weekend travel.
- **Duration within day**—Most travel forecasts are for a typical 24-hr period. These can be adjusted to simulate peak-hour and peak-period conditions with time-of-day factors. Air-quality models typically need emissions for each hour of the day. This task is often performed using travel

start times by hour and trip type from travel diaries. However, these adjustments are highly uncertain and rarely validated.

- **Travel by grid**—Most travel forecasts are conducted for specific transportation facilities or segments. However, air-quality models often need information by "grid," or aggregations of transportation facilities. Geographical information system software is an efficient tool for aggregating travel into grids.

Multidimensional Synergistic Impacts from Adjustments to Travel Activity Results

As needs for precise estimates of emissions and air quality grow, it is common to use travel activity inputs based on typical weekday travel that may have been adjusted for season, weekend travel, time of day, and grid. Demands for even greater precision might require travel activity estimates for specific vehicle age categories, meteorological conditions, or vehicle types. The committee feels that the level of detail associated with current travel-demand models is insufficient to make these simultaneous adjustments without introducing substantial additional uncertainty. A good example of this problem is the need in air quality modeling to estimate aggregate heavy-duty vehicle (HDV) activity and adjust these estimates for time of week and time of day. Because HDVs produce a disproportionate amount of NO_x emissions, they greatly impact the ability to model ozone accurately. Yet the multiple adjustments of travel activity results needed to produce estimates of HDV activity by time of day introduces an unknown level of uncertainty to emissions and air quality simulations.

Emissions Modeling

A primary use of MOBILE is for developing on-road mobile-source emissions inventories for use in air-quality planning. Emissions rates developed in MOBILE are combined with average vehicle speeds and travel activity estimates to develop these inventories. The emissions rates generated by MOBILE require a multitude of input assumptions. For most input assumptions, MOBILE provides national default values or users can input locally specific values. Regions are free to choose when to use national defaults or local data. This decision is sometimes made on the basis of whether the national default or local data will positively affect their attainment demonstration or conformity analysis. MOBILE is particularly sensitive to input assumptions for vehicle age, VMT by vehicle class, average vehicle speeds, and temperature—all of which can vary widely from region to region. Below is a brief discussion of parameters that are important in the application of MOBILE to an individual region. Chapter 3 of the report discusses the technical components of MOBILE in more detail.

Vehicle Registration

MOBILE uses vehicle registration data to determine the percentage of the vehicle fleet for each combination of vehicle type with vehicle age. MOBILE combines this information with average mileage accumulation rates to determine the fraction of overall travel in a region associated with each vehicle type disaggregated by age and average fleet emissions rates. The MOBILE documentation published by the U.S. Environmental Protection Agency (EPA) strongly encourages users of MOBILE to develop locally specific vehicle registration distributions because the default values reflect national averages for 1990.

Emissions inventory estimates are affected by assumptions about vehicle registration distributions. Within an urban area, the vehicle fleet composition can vary significantly across subregions, for example, in relation to development patterns and the economic status of the population. Areas with newer development and higher average income levels tend to have newer vehicle fleets, resulting in lower emissions rates than in older areas. The choice of a particular vehicle registration distribution can affect on-road emissions inventories by approximately 5 to 10% (Pollack et al. 1991). As a result, estimates of on-road mobile-source emissions require accurate vehicle registration distributions at an appropriate level of detail for a particular application.

Vehicle Miles of VMT Travel Mix

A VMT mix identifies the percentage of VMT that is accumulated by each of the eight vehicle classifications used by the MOBILE model. MOBILE uses this VMT mix to generate composite vehicle-emissions factors. MOBILE calculates a typical urban area VMT mix based on national data for several variables, including registration distributions, annual mileage accumulation rates, percentage of diesel sales, and number of vehicles. EPA recommends that users develop locally specific estimates of VMT mix for SIP emissions inventories.

Policy decisions regarding mobile-source controls are affected by assumptions incorporated into the VMT mix data. National default values based on averages might not accurately represent the VMT mix in any given region. For example, if a particular region has a high percentage of heavy-duty vehicles, the default VMT mix would lead to an underestimation of NO_x emissions.

Average Speed

MOBILE uses regional average vehicle speeds estimated by travel demand models to develop emissions rates. The model develops base emissions rates for various vehicle classes using standard driving cycles such as the Federal Test Procedure (FTP). These base emissions rates are then adjusted to a particular location's average speed using speed correction factors. Speed correction factors are intended to reflect the differences between emissions rates under test conditions and emissions rates under regional driving conditions. More detailed descriptions of the use of these test cycles, speed correction factors, and facility correction factors are contained in Chapter 3.

It is important to point out that traditional travel-demand models do not estimate average speeds directly, but rather produce average speeds from estimates of traffic volumes. This is important because these speed/ volume relationships are not very accurate and are sometimes adjusted during calibration so that modeled traffic volumes match observed volumes.

Temperature

MOBILE requires locally specific temperature data, in part because no national defaults would be appropriate for temperature. Thus, users must develop average temperature data to develop on-road mobile-source emissions inventories. Analysis of the MOBILE model shows that at higher temperatures, a one-degree change in temperature results in a 1% change in emissions factors (Pollack et al 1991), though this premise has not been validated. As a result, temperature variations within a region may significantly impact emissions inventory estimates.

Air-Quality Modeling

Air-quality models have become the central tool for analyzing how future emissions changes, including changes due to new control strategies, will affect air-quality (NRC 1991). Ozone, for example, is not produced directly from emissions sources, but rather through complex chemical reactions involving VOCs, NO_x, and sunlight (solar radiation). High ozone episodes occur during periods of stable atmospheric conditions that are accompanied by high temperature and low winds. Regional photochemical models attempt to predict the formation of ozone for multi-day events using meteorological data, emissions inventory data, and complex air-chemistry equations. In contrast, CO problems are much more localized in nature and require different air-quality models that represents site-specific dispersion of this pollutant. Vehicles emit CO directly, and its local concentration depends upon the rate that it is emitted and dispersed in the atmosphere. A major problem for air-quality managers is identifying controls that will reduce CO, ozone, and PM. On-road mobile-source emissions are important contributors to each of these pollutants. For most of the nation, the estimation of on-road emissions data needed for air-quality modeling studies are derived from MOBILE using results from travel-demand modeling.

In a typical application (e.g., analyzing ozone-control strategies), an air-quality model will be applied to simulate photochemical pollutant concentrations during a 3- to 5-day episode, using emissions and meteorology data specific to the period of application. Typical model resolution, and hence the scale of emissions inputs, is approximately 5 kilometers (km). The modeling results are evaluated against observed ambient measurements to assess the validity for use in control-strategy assessment. Errors in the emissions and other model inputs are evident from disagreements between observations and model simulations, not only for ozone, but for the precursors as well. Because of nonlinearities in the formation of ozone, it is important that the observations and the simulated values agree reasonably well. Otherwise, the model response to emissions changes will be suspect. With this in mind, EPA has developed model performance guidelines (EPA 1999b).

At present, it appears that uncertainties in the emissions, and mobile-source emissions in particular (NRC 1991; Harley et al. 1993a, b), are major contributors to poor model performance. As described in more detail in Chapter 4, a variety of studies have concluded that mobile-source VOC emissions are significantly underestimated in the models. The performance of air-quality models improves significantly when estimates of mobile-source VOC emissions are increased. However, there are many other factors that contribute to the poor performance of air quality models, including errors in the atmospheric chemical mechanisms and meteorological inputs within the air quality models as well as uncertainties in the travel activity inputs discussed earlier. After a model achieves acceptable performance, it is used to test control strategies, in particular to identify the set of controls that are likely to lead to attainment. Historically, attainment has meant that predicted ozone is less than 0.12 parts per million (ppm) in each subarea of the region. Thus, the target is an absolute number, and modeling results are used in an absolute sense.

The use of MOBILE to develop emissions inventories for ozone modeling (and PM modeling in the future) highlights the evolution of the demands placed upon it. Such applications require estimates at relatively fine spatial and temporal resolutions, not simply across a fleet of vehicles in a broad urban area. Topographical features, such as hills, might play a major role in emissions and atmospheric processes. Further, the varying composition of the fleet can become important if one area of a city is more likely to contain high-emissions vehicles than another. MOBILE was not originally designed to support applications that require a high resolution and accuracy of emissions inventories. It is not apparent to the committee that an emissions-modeling tool designed specifically for use in air-quality modeling would have been designed in the same way as MOBILE.

The importance of providing finer spatial resolution is highlighted by a recent study (Lackshminarayanan, 1999). That study used a mobile source inventory for the Atlanta, GA, area, as developed by MOBILE5a, to simulate ozone, nitrogen dioxide and an air toxic (formaldehyde) concentrations in the region. Next, results from MEASURE (Guensler et al., 1998; see Chapter 5 for details) were used to spatially and temporally real-locate those emissions, thus keeping the same basin-wide mass emissions of each species, but changing the details of the time and location of emissions. In this process, the study was also able to develop the emissions at finer grid resolutions than the 4×4 km MOBILE inventory. The photochemical model was then re-applied using grid resolutions of 1×1 km, 2× 2 km and 4×4 km. Peak levels of nitrogen dioxide and formaldehyde were found to be very sensitive to grid size, varying by up to a factor of five or more. Ozone levels were less sensitive. Thus, this study concluded that accurate exposure assessment of primary pollutants, such as air toxics and particulate matter, are likely to require spatially and temporally detailed emissions information.

In part because of the difficulties in estimating emissions inventories, guidelines for future ozone air-quality modeling might be used in a more relative sense. Thus, rather than ensuring that all concentrations simulated by the model are at or below 0.12 ppm (or 0.08 ppm for the new 8-hr standard), a relative reduction from a base scenario could be used to assess the adequacy of emissions controls (EPA 1999b). For example, if the base calculation led to a maximum ozone concentration of 0.156 ppm, and the control case had a peak of 0.132 ppm, this would represent a reduction of 15%. That 15% relative reduction would then be used to test for attainment. If the design value was 15% over the limit, then the modeling test would be passed.[1] The use of the model in such a relative sense is believed to be more accommodating of uncertainties, such as those in the emissions

[1]For a more detailed explanation of this type of attainment demonstration, see EPA (1999b).

themselves. Thus, as MOBILE has been evolving to provide absolute emissions levels (as opposed to relative levels), the regulatory application of air-quality models may be moving towards relative uses, in part, because of problems in validating these models.

Ozone modeling is not the only purpose for a model such as MOBILE. Air toxic agents are a societal concern, and automotive emissions contain significant quantities of hazardous air pollutants (HAPs) or air toxics, such as benzene, formaldehyde, and 1,3 butadiene. Unlike ozone, elevated levels from these primary emissions are found near roadways. Concentrations in nearby areas might be much reduced by dispersion. A recent California Air Resources Board (CARB) study concluded that concentrations of pollutants inside vehicles can be two-to-seven times greater than concentrations at air-monitoring stations (CARB 1998). This has important implications in exposure assessments, because it places greater emphasis on knowing the detailed spatial location of emissions in relationship to potentially exposed populations. Unlike ozone modeling, which might require emissions with a spatial resolution of 4 km or so, HAP exposure assessment might require resolution at a scale of tens or hundreds of meters. MOBTOX (or a version of MOBILE that estimates air toxic agents), in principle, could be applied at this level, but it lacks many features that might be important at such fine spatial scales (e.g., the influence of topographical features and specific traffic-control measures) that get averaged out over larger areas. The use of MOBILE to model the concentrations of HAPs near roadways raises issues similar to those of using MOBILE to model micro-area carbon monoxide.

Users of Modeling Components

Both the public and private sectors use the transportation, emissions-factor, and air-quality modeling components of air-quality planning and regulatory processes. The broad community of model users includes government agencies, private consulting firms, public interest groups, and other researchers. The primary purpose for this modeling effort is to fulfill specific transportation and environmental legislative and regulatory requirements.

Governmental users include agencies at the federal, state, regional, and local levels that use these models to conduct transportation and environmental analysis. Private consulting firms often contract with government agencies and industry to conduct specific planning and environmental studies using these modeling tools. Public interest groups, such as environmental organizations, are often stakeholders in transportation and environmental planning studies that rely on these technical-modeling tools. Universities train future users of these models, and they perform research and data-collection activities that support the continued development of these tools.

Level of Analysis and Model Uses and Users

There are four distinct classes of uses of this compendium of models in air-quality management, as described below and summarized in Figures 2– 2 through 2–5:

- **Level 1**—Direct use of the MOBILE model to estimate emissions rates. This is often used by EPA to assess national mobile-source regulatory strategies (Figure 2–2).
- **Level 2**—Use of transportation and MOBILE models to estimate emissions inventories. Such analyses are used by state environmental agencies, transportation consultants, metropolitan planning organizations, and state departments of transportation to support development of SIPs and transportation conformity analysis (Figure 2–3). Such analyses are also used by EPA to estimate the impacts of national mobile-source regulatory strategies on overall emissions levels.
- **Level 3**—Use of transportation, MOBILE, and air-quality models to simulate pollutant concentrations. This is usually done by state environmental and transportation agencies and universities to assess attainment of National Ambient Air Quality Standards (NAAQS) (Figure 2–4).
- **Level 4**—Use of transportation, MOBILE, air-quality, and exposure models to simulate human exposure and health impacts from air pollutants. This is done by health professionals at universities and environmental agencies to help assess mortality and morbidity associated with air pollution (Figure 2–5).

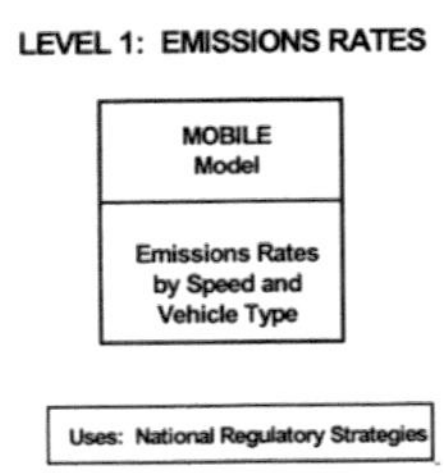

FIGURE 2–2 Use of MOBILE for estimating vehicle-emissions factors.

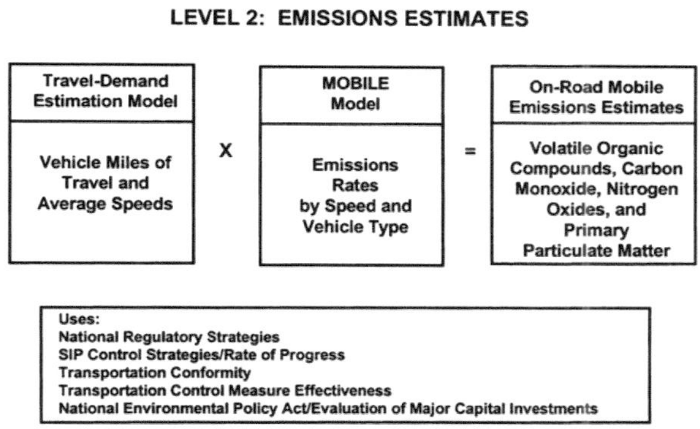

FIGURE 2–3 Use of MOBILE for estimating vehicle emissions.

Fidelity, Accuracy, and Precision of Each Component

Fidelity, accuracy, and precision all relate to the ability of a model to simulate the real world. Table 2–1 provides a brief summary of how each of these modeling fields approaches calibration and validation. It emphasizes that travel-demand models have standard calibration and validation statistical performance metrics. It is common in travel-demand modeling to compare predicted traffic volumes on a roadway segment or ridership on a transit line with independently observed values. Such standard calibration and validation procedures and metrics have not been developed for applications of MOBILE. It should be noted, however, that regions rarely validate travel-demand models to ensure that the underlying behavioral assumptions are accurate and that a model predicts the right volumes for the right reasons over time and across the region.

Table 2–1 also shows that, although air-quality models are critical for policy implementation, they have greater tolerances for gross error. Besides errors inherent to air-quality modeling itself, this is also due to the propagation of errors that occurs through the linkage of models. Errors that occur in the travel-demand estimates propagate and add to the errors in the mobile-source emissions model, which in turn, propagate and add to the errors in the air-quality models.

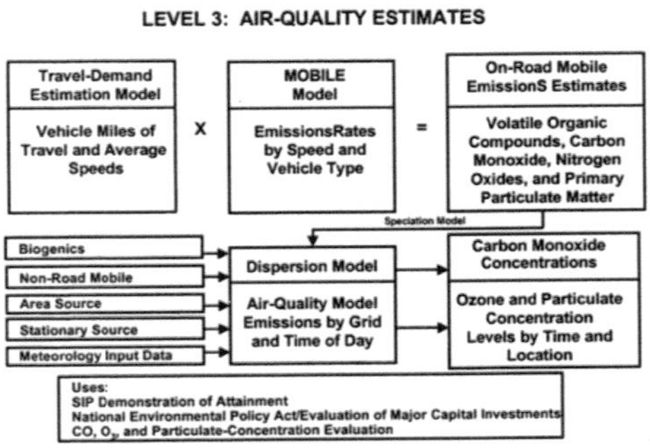

FIGURE 2–4 Use of MOBILE in the estimation of ambient pollutant concentrations.

USES OF MOBILE IN POLICY DECISION-MAKING

The following section summarizes the uses of MOBILE, including a discussion of who is responsible for executing MOBILE and assessing the model's results, modeling issues particular to the application, and future directions EPA should consider. Although the six application areas vary widely in terms of their spatial and temporal scales, the same MOBILE model is used in all settings. These descriptions further illustrate that MOBILE is currently applied in ways for which it was not developed and for uses beyond which it is well suited. EPA has been urged to develop other models more suited to specific purposes. The next version of MOBILE, MOBILE6, is intended to narrow this gap, yet the discussion below suggests that the MOBILE structure itself cannot adequately serve such a wide range of uses.

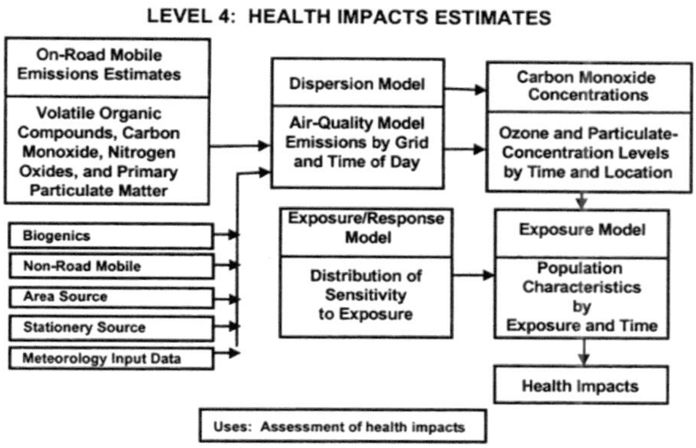

FIGURE 2–5 Use of MOBILE in assessing health impacts.

National and Regional Regulatory Strategies

Primary Users and Purpose

EPA uses MOBILE to evaluate a variety of national air-quality regulatory strategies, including new-vehicle emissions standards, fuel-quality specifications, and inspection and maintenance (I/M). EPA also uses MOBILE for evaluating the contribution of on-road vehicles to the nation's total air-pollution emissions and to inventory and monitor historical trends (EPA 1998a; EPA 1998b).

EPA's most significant national vehicle emissions-control regulatory effort may relate to concurrent adoption of Tier 2 new-vehicle emissions standards and new limitations on sulfur in gasoline. Because of the significance of this action, EPA has undertaken a major effort to develop a version of the model that is as close as possible to the expected new MOBILE6 to assess this major regulatory development (EPA 1999c). The version of the model is known as the Tier 2 model, which is a spreadsheet model incorporating elements of MOBILE5b and MOBILE6.

The Tier 2 motor-vehicle emissions standards (EPA 1999a) involve reducing new-vehicle emissions standards beginning in 2004. By 2007 pas senger car and light-duty truck emissions standards will be lowered to an average of .07 grams per mile (g/mi) for NO_x. The final rule for the Tier 2/Sulfur Gasoline Program also requires that gasoline produced by refiners or sold by importers meet an average sulfur content of less than or equal to 30 ppm by 2007. Figure 2–6 shows the impact of these new requirements generated by the Tier 2 version of MOBILE (EPA 1999d). In this figure, passenger cars are labeled LDV and trucks are labeled LDT1/2 (light pickups, minivans, and most sport utility vehicles) or LDT3/4 (heavier pickups and sports utility vehicles). This figure shows very large decreases in NO_x and PM-2.5 emissions from the combination of Tier 2 and sulfur controls. Emissions of NO_x and PM-2.5 in year 2020 are estimated to be cut by two-thirds from projected emissions without these controls. It should be noted, however, that these standards must withstand any legal challenges.

TABLE 2–1 Comparison of Calibration and Validation Standards for Travel-Demand and Air Quality Models with the MOBILE Model

Model Structure Evaluation	Travel-Demand Estimation Model	MOBILE	Air-Quality Model
Typical calibration method	Household travel behavior survey with sample size specification (90% confidence, 5% error	Laboratory estimates of emission rates (under-represented sample size, no explicit performance standard)	Laboratory estimates of precursor sensitivity to ozone formation
Validity to second data source	Field test of passenger volume estimates (traffic counts and transit ridership) by segment (88% correlation, 5% error)	Field test of emissions levels (e.g., tunnel tests, no explicit performance standard)	Field test to ozone monitor-selected episodes (30% gross error ±15% bias)

Sources: Pederson amd Samdahl 1982; EPA 1999b.

Issues and Limitations

MOBILE is best suited for national or regional applications because it utilizes an aggregate approach appropriate for wide areas and long time-

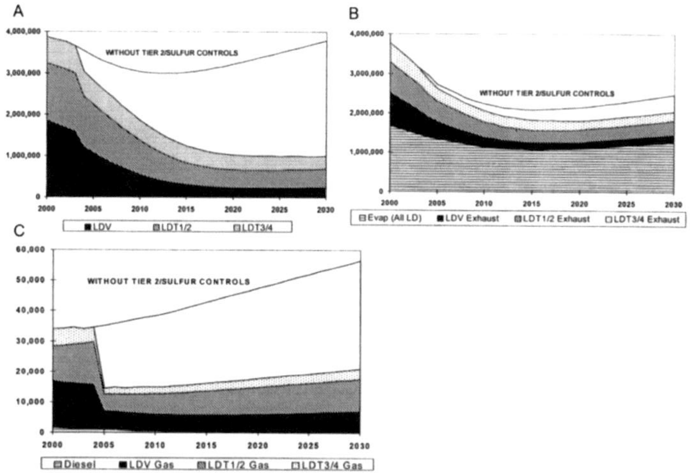

FIGURE 2–6 Projected annual vehicle emissions (in annual tons) with Tier 2/ fuel sulfur reductions compare to projected emissions without these controls. 47-state light-duty vehicle emissions of (A) NO_x, (B) VOCs, and (C) PM2.5 with Tier 2 vehicle emissions standards and reductions in fuel sulfur content. Emissions are in tons. Passenger cars are labeled LDV and trucks are labeled LDT1/2 (light pickups, minivans, and most sport utility vehicles) or LDT3/4 (heavier pickups and sport utility vehicles). Also labeled are diesel passenger cars and diesel light-duty trucks. Source: EPA 1999d.

scales under average conditions. In regional applications, however, significant error can be introduced using MOBILE's generic national defaults.

Policy Implications and Future Direction

Output from MOBILE for use in national and regional air-quality strategies has profound technical and economic effects on the direction of the nation's air-quality management activities. Significant inaccuracies can result in misdirected control strategies that prolong public exposure to health hazards or waste large sums of money. It is not certain what major inaccuracies will exist in the new MOBILE6, although clearly the major changes from MOBILE5b indicate that there have been such inaccuracies in the past.

Evaluation of Control Strategies, Emissions Inventory, and Rate of Progress

Primary Users and Purpose

State and local governments, MPOs, consultants, research institutions, and others apply MOBILE to develop regional emissions inventories, evaluate alternative mobile-source emissions-control strategies, and track trends in control-strategy implementation. In particular, MOBILE is used in ozone nonattainment to demonstrate how a region will comply with the Clean Air Act Amendment of 1990 requirement to reduce VOC emissions by 15% from 1990 to 1996 and 3% annually until attainment.

Issues and Limitations

Some applications, particularly region-wide ones, warrant an aggregate approach to estimating vehicle emissions. In such applications, it may be adequate to develop representative emissions factors from region-wide parameters, such as average vehicle speeds by facility class (see Chapter 3 for a description of facility correct factors) and registration distribution, that can be combined with VMT to estimate regional emissions. MOBILE is well-suited to these types of applications. However, there are some concerns about the inability to see all the underlying assumptions in the model, some of which might have to be altered to fit local observations. The committee feels that such aggregations still introduce inaccuracies in the end result because the assumed average may not truly represent the real average in a skewed distribution. Some regional-control strategies cannot be modeled within the current MOBILE structure, including land use changes, vehicle scrappage, and clean-fueled fleet incentive programs.

The accuracy of projections used in SIP development depend to a great extent on the accuracy of assumptions about such factors as fleet turnover, effectiveness of control strategies, and future deterioration rates of vehicles. Generally, there is concern that national default parameters coded in MOBILE5 are outdated, and that their use in SIP development is not appropriate. Additionally, overall concern is high that the vehicle-operating patterns in the model do not correspond to present and future on-road conditions. EPA is addressing some of these concerns through the addition of an "off cycle" factor that accounts for higher average speeds and air conditioning use. Improving these aspects of MOBILE will also require improvement in vehicle-activity.

Policy Implications and Future Directions

MOBILE users have many questions about its accuracy. For example, MOBILE's ability to correctly evaluate the impacts of air quality improvement initiatives, such as vehicle emission inspection and maintenance programs (Harrington et al. 1998) and the use of oxygenates in winter (NRC 1996; NSTC 1997) has been questioned. The development of SIPs requires accuracy in emissions inventories and crediting of emissions reductions from controls, both of which are particularly sensitive to errors. Little has been done to address this issue, and it will undoubtedly become more significant in the absence of a significant model revision based on better data and science. To date, MOBILE revisions have been infrequent, with the last major update (MOBILE5) released in 1993. It is unclear at this point whether the upcoming version of the model, MOBILE6, will increase or decrease regional emissions predictions compared to MOBILE5. It is likely that at least VOC emissions will increase. This uncertainty as well as delays in the release of MOBILE6 greatly complicate SIP development.

A long-range plan is needed to determine the appropriate update frequency schedule and whether there should be updates issued more continuously. There is significant concern among the committee about the seven-year gap between MOBILE5 and MOBILE6. This delay results in the use of a version of MOBILE containing information known to be obsolete and incorrect. Users need updates that incorporate the latest findings on the factors that affect emissions and the effectiveness of control strategies so that SIPs can be based on the most accurate information. One possibility may be to allow users access to new information and allow them the flexibility of incorporating such information into SIPs and other planning pro cesses. However, there are also problems that might come with more frequent or continuous updates, such as inconsistencies between models used in SIP budgets and subsequent conformity determinations.

The need for more accurate emissions inventories and assessment of controls requires expanding the capabilities of MOBILE or developing new models. Users desire more disaggregation of model inputs so that local conditions can be better represented. There is a need to assess impacts of some alternative strategies that are not presently incorporated in MOBILE, such as alternative-fueled vehicles. Road grade is an important local factor that will not be incorporated in the current or updated versions of MOBILE. Users will also expect that some form of instantaneous modeling will be provided so that both individual projects and larger-area transportation systems can be assessed in a more accurate manner.

SIP Demonstration of Attainment

Primary Users and Purpose

States and, in some cases, metropolitan planning organizations (MPOs) are required to develop a demonstration of attainment for SIPs in ozone and CO nonattainment regions. These must be submitted to and approved by EPA. Such an analysis demonstrates that the proposed emissions-reduction strategies will attain and maintain ambient ozone and CO concentrations below the NAAQS. For this, an urban or regional-scale air-quality model is used with on-road mobile-source emissions estimated from MOBILE and data on other emissions sources (stationary, biogenic, area sources, and non-road mobile sources).

Issues and Limitations

Regional air-quality models are complex and generally require detailed temporal and spatial allocation of emissions. Applications of urban and regional air-quality models usually simulate a 3- to 10-day ozone episode. The model requires hourly gridded emissions for NO_x and VOCs, as well as speciation of VOCs. Additionally, emissions must be disaggregated by emissions mode (e.g., exhaust and evaporative), technology, and emitter group. As discussed in Chapter 4, there is strong evidence that MOBILE emissions, including emissions by mode, are inaccurate and thus inhibit accurate air-quality modeling. Inconsistencies may also result from other emissions sources, due to the lack of temporal and spatial detail needed to model regional air quality. For example, locomotive, marine, and construction emissions might be difficult to accurately characterize because critical data are proprietary.

Policy Implications and Future Directions

One important output from MOBILE is emissions-rate detail for inclusion in SIP photochemical modeling. Ultimately, photochemical model accuracy depends on the accuracy of the emissions estimates, as well as the methods used to spatially and temporally allocate on-road emissions estimates, and to speciate the VOC emissions. Errors in these steps can potentially lead to inaccurate conclusions and selection of sub-optimal control strategies. Because of difficulties in accurately determining overall emissions inventories, future guidelines for demonstrating attainment may be based on a relative rather than an absolute reduction in maximum ozone concentrations. This, however, will not eliminate the issues associated with spatially and temporally allocating emissions within a region. In addition, a lower NAAQS standard for ozone will require better spatial and temporal disaggregation over wider regions.

Transportation Conformity and Evaluation of Transportation Impacts in a Nonattainment Area

Primary Users and Purpose

The MPO is responsible for performing an air-quality conformity analysis for nonattainment areas. Conformity is a determination that emissions from transportation plans, programs, and projects in a nonattainment area do not exceed mobile source emissions budgets established in SIPs (Federal Highways Administration 1992). The conformity demonstration is intended to show that transportation activities will not cause or contribute to any new violation of air-quality standards, increase the frequency or severity of existing violations, or delay timely attainment of standards (EPA 1993a). The conformity analysis is done for the system of projects contained in a region's transportation improvement program (TIP) and transportation plan (see glossary for definitions of transportation improvement program and transportation plan). Regions in CO and PM-10 nonattainment must also conduct project-level conformity analysis ("hot spot" analysis) for critical intersections and sites with violations or possible violations of the NAAQS.

Similar to emissions inventories developed for a SIP, the transportation conformity analysis consists of determining emissions estimates as a function of vehicle activity and region-specific emissions factors. However, this must be done for a specific system of projects or programs, and on a smaller scale than for the SIP. In some areas, where planning organizations have limited responsibilities, the state departments of transportation may perform the technical analyses. The purpose of an air-quality conformity analysis is to ensure that transportation projects identified in a region's long-range transportation plan and short-range TIP are consistent with local air-quality goals and objectives. The conformity analysis is needed to ensure that calculated emissions levels for federally funded and regionally significant projects do not cause emissions to exceed the targets specified in the relevant SIP.

Issues and Limitations

A critical aspect of the conformity determination is that it is developed for a system of individual projects. Because MOBILE uses top-down procedures to estimate emissions, it is not suited for conformity applications. The need for microscale modeling (the modeling of specific corridors or intersections) in the conformity analysis is not adequately supported by the aggregate regional and national data and assumptions used in MOBILE. The conformity analysis also must show consistency among the SIP, transportation plan, and TIP. This task can be challenging because the transportation plan horizon date is much further out than that of the SIP. The SIP is based on the attainment dates set in the Clean Air Act and the transportation plan horizon is 20 years. Federal conformity regulations require analysis for the "out years", the years beyond the deadline of the SIP but within the deadline of the long-range transportation plan. This creates an inconsistency problem because the maximum growth a region can accept depends on the level of vehicle technology and fuels that are accounted for in the SIP.

The current use of MOBILE5b model in conformity analysis is limited by the model's ability to appropriately model emissions estimates for the out years beyond the SIP deadline. Specific out-year technological assumptions used when MOBILE5 was developed in 1993 may not accurately represent current assumptions. This could cause an unnecessary strain on regions, as they are forced to meet transportation plan budget tests based on outdated forecasts.

For example, prior to the regulations implementing the national low-emissions vehicle (NLEV) standards and new regulations on heavy-duty diesel vehicles (HDDVs) in 2004, a region experiencing rapid growth would find it difficult to pass a transportation plan horizon-year budget test due to vehicle activity outpacing vehicle technology. Figures 2–7 and 2–8 demonstrate this observation using conformity analysis results from a nonattainment area (North Central Texas Council of Governments 1998). Estimates that include the effects of the new NLEV and HDDV programs in MOBILE, show that NO_x and VOC emissions will be reduced by 43%

and 32%, respectively, by the year 2020. Since this time, EPA has developed even more dramatic improvements through Tier 2 emissions standards and fuel sulfur reductions.

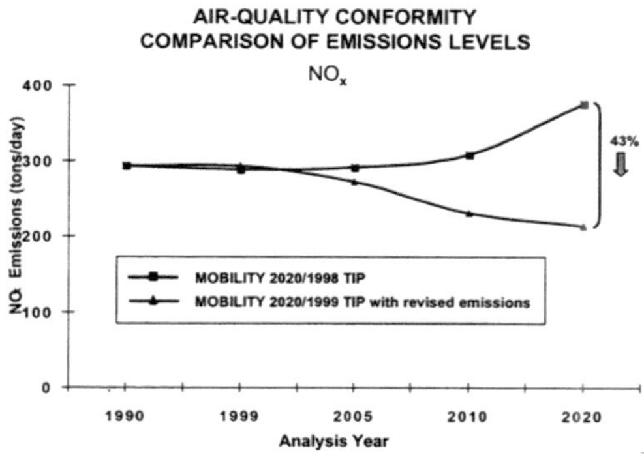

FIGURE 2–7 Effect of NLEV program and new HDDV regulations on NO_x emissions for Dallas metropolitan region. Source: NCTCOG 1998.

Policy Implication and Future Direction

Conformity analysis is often conducted annually for the TIP development and every three years for the transportation plan. This analysis is most critical to MPOs and state transportation departments because of its potential impacts on constructing transportation projects. In many cases, the conformity determination is made on a system of projects that show relatively small differences in emissions, especially compared to the effects of vehicle technology assumptions. MOBILE's aggregate approach to emissions estimates makes it poorly suited to accurately characterize such relatively small emissions impacts. Additionally, transportation projects often have a complex impact on traffic characteristics, and hence emissions, that are difficult to represent in the current linkage of travel-demand models to MOBILE. It is important that MOBILE be improved so that it is able to more accurately perform conformity or that EPA develops tools especially for such analysis.

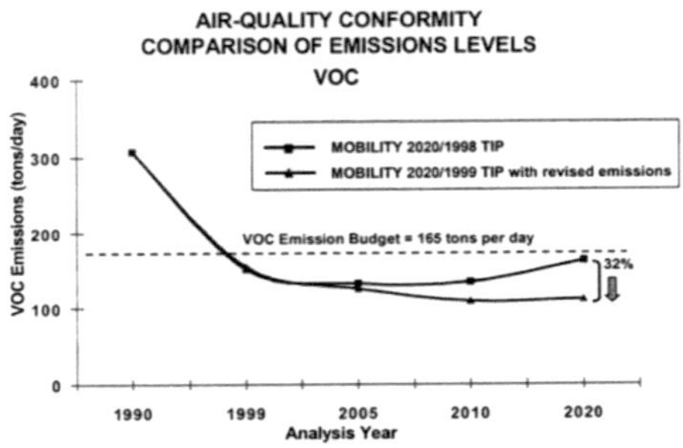

FIGURE 2–8 Effect of NLEV program and new HDDV regulations on VOC emissions for Dallas metropolitan region. Source: NCTOG 1998.

Conformity analysis will require that either MOBILE or an alternate conformity tool be quickly adaptable to vehicle technology advances and future regulatory initiatives. Examples include accounting for the proportion of the vehicle fleet to be zero-emission vehicles (ZEV), the Tier 2 emissions standard, and limits on sulfur in gasoline. Although these modifications are part of the model used for assessing the Tier 2 proposal (the Tier 2 Model), and will be a part of MOBILE6, they cannot be accounted for in MOBILE5b, the model currently used by state and local agencies for SIP development and conformity analysis. Due to the timeline inconsistency between SIP and transportation plans, this is most relevant for a transportation conformity analysis, because it is these new technologies and initiatives that have the largest impact on out-year emissions.

Transportation Control Measure Effectiveness and CMAQ Eligibility

Primary Users and Purpose

States and MPOs use both travel-demand models and the MOBILE model to aid in the selection of transportation-control measures (TCM) for

SIP credit and to aid in the evaluation and selection of appropriate projects for congestion mitigation and air quality (CMAQ) funding. TCMs are projects specifically designed to assist in reducing overall emissions by reducing travel demand (VMT and trips), encouraging the use of alternative modes, and smoothing traffic flow. Traditional TCMs include high-occupancy vehicle lanes (HOV), signal and intersection improvements, bicycle and pedestrian improvements, light or commuter rail, advanced transportation management technology (e.g., freeway management of incidents and accidents), and travel-demand management strategies (e.g., parking pricing). TCMs are evaluated and selected for CMAQ funding often using technical methodologies to estimate their effects on emissions. After evaluation and commitment by the MPO, these projects can be inventoried and used in emissions-reduction strategies in SIPs.

Issues and Limitations

One critical issue with TCM and CMAQ projects is the inability to evaluate their impacts using the traditional travel-demand modeling process outlined in Figure 2–1. Travel-demand models, the backbone of transportation planning, cannot assess small-scale project specifics such as those commonly found with intersection and signal improvements. Therefore, air-quality impacts of most TCM and CMAQ projects are evaluated "off-model" or with post-processing techniques and do not benefit from the many internal travel-model features affecting volumes and speeds regionwide. As with the conformity analysis, this affects the accuracy of the emissions-reduction estimates from MOBILE because MOBILE is used to estimate the effects of a small change in travel parameters on a subset of the overall vehicle fleet.

Because no formal blueprint outlines appropriate methodologies at the national level, each region has developed its own approach to evaluating TCM and CMAQ priorities, and these variations may cause important nationwide inconsistencies. Consistency in evaluation of TCM and CMAQ projects has been enhanced by post-processing software packages that estimate several useful measures and assess the likely effects of a particular project. Such models are designed to link directly to the traditional four-step travel-demand modeling process through trip tables, which are passed back and fourth as necessary. Unfortunately, software packages to estimate effects on travel activity or air quality do not exist for all TCM and CMAQ categories, and regions must devise their own methods to evaluate some TCMs (e.g., intelligent transportation systems for freeway management).

Policy Implications and Future Directions

Nonattainment regions must evaluate TCM and CMAQ project categories, and these areas would benefit from evaluation methods that quantify changes in nontraditional transportation and in air quality. This will improve consistency, accuracy, and efficiency, for which national benefits could be inventoried. As with conformity analysis, the current MOBILE model is poorly suited to estimate the emissions-reduction benefits of TCMs. Users need more refined tools that provide a greater resolution of the impact of TCMs on traffic flow and emissions.

National Environmental Policy Act and Evaluation of Major Capital Investments (Transit and Highway)

Primary Users and Purpose

The National Environmental Policy Act (NEPA) requires documentation of the environmental impacts caused by major capital investments that use federal funds, such as the construction of major transit and highway projects. NEPA requires that a project will not result in a violation of air quality standards and that the project be included in a TIP. NEPA also requires planners to provide a relative comparison of the air quality impacts of alternatives including the no-build alternative. Many agencies rely on MOBILE results to evaluate air-quality impacts and suggest alternative transportation investment options. The primary users of the MOBILE model in evaluating major capital transportation investments and developing environmental impact studies are state departments of transportation, federal resource agencies, state resource agencies, MPOs, local governments, consultants (generally working for these government units), and universities.

Issues and Limitations

MOBILE is designed to evaluate emissions impacts on a regional level not at finer levels such as corridors. Because many of the internal defaults in MOBILE are based on national and regionwide estimates, it cannot provide the resolution needed to assess impacts for individual corridor-specific projects. One example to demonstrate this point is that the vehicle registration data used to estimate vehicle age at the county level is not the same in each corridor in the county, especially considering the dramatic variations in income levels within a county.

Policy Implications and Future Directions

Because MOBILE is the recognized national emissions factor model, its results have broad impacts. For example, it may be used to estimate the difference between a four-lane and a six-lane freeway, the effects of a new rail line or a reversible HOV lane, or the impact of a range of TCM strategies. As described above, MOBILE is clearly not the most appropriately designed model to carry out such analysis. It is recommended that another tool be developed to accurately quantify small-scale impacts. However, EPA must first approve a new model before it is used to evaluate emissions impacts. Thus, not only does there need to be development of improved methods for evaluating such projects, the EPA will also need procedures for vetting, documenting, and evaluating new models and methods for estimating corridor-level and other small-scale emissions impacts.

SUMMARY OF POLICY IMPLICATIONS AND RECOMMENDATIONS

Modeling the air-quality impacts of on-road vehicles is an interdisciplinary effort that encompasses the modeling of travel demand, emissions, and air quality. These individual components must be systematically integrated to develop analyses with adequate consistency, fidelity, accuracy, and precision. Each of the three types of models has a different focus (i.e., transportation models generally focus on transportation segments, emissions models on engine modes, and air-quality models on photochemical reactions and dispersion). Some kinds of uncertainties are inherent within each modeling domain, others occur at the interfaces when output from one model is used as the basis for input into the next. EPA should take steps to improve the linkages among the three models and improve the methods that are used to process MOBILE outputs for use in regional air-quality modeling.

MOBILE is a single piece of software with a minimum of six different categories of uses. It is best suited for aggregate analysis and assessment of national and regional regulatory strategies and the development of SIPs for metropolitan areas. It is poorly-suited for analyses of a system of projects or corridor analyses characteristic of conformity applications, assessment of TCMs, and environmental-impact assessments.

Inconsistencies among these differing categories of uses led the committee to conclude that likely no single model is appropriate for all applications. As described further in Chapter 6, the use of MOBILE should be supplemented with the development and adoption of alternative models specifically designed to better link traffic flow in local settings to emis sions. EPA should develop one or more new mobile-source emissions modeling processes, perhaps using a Geographic Information System platform, that incorporate localized driving cycles and other conditions that influence emissions. EPA should also consider the development of partnerships for data collection to better characterize emissions rates under a variety of conditions.

MOBILE has a critical role in estimating and managing the levels of mobile-source emissions control. Future focus on new emissions standards, the growing concern about air toxic emissions, and the growing cost of control strategies will increase the demands for accuracy and detail from MOBILE. A strategic and comprehensive long-range plan is needed to better identify emerging needs for the modeling of automotive emissions, define the levels of detail and accuracy needed to meet those needs, and set priorities to improve MOBILE.

EPA should modify its policy of issuing MOBILE updates on a batch, infrequent basis. Providing updates (known as information fact sheets) for major factors as soon as they are known, such as the adoption of NLEV standards or emissions-reduction credits for oxygenated fuels, would allow users to account for the latest technologies and revisions in SIP development as soon as possible. The committee recognizes the difficulties of working with a moving target, but concludes that the benefits of up-to-date modeling outweigh the disadvantages of more frequent changes.

3

Technical Issues Associated With the MOBILE Model

HE FOCUS OF CHAPTER 3 is the technical issues associated with the current (MOBILE5) and upcoming (MOBILE6) versions of the MOBILE model. The introductory portion of this chapter discusses the development of the model and the updates for MOBILE6. The chapter goes on to describe related models for estimating mobile-source emissions (PART5, NONROAD, and others) and previous reviews of the model. A major portion of this chapter then focuses on the technical issues associated with the model, such as how the model handles high emitters, driving cycle, start emissions, and many others details. The chapter concludes with a summary and recommendations related to technical aspects of the model.

HISTORY AND STATUS OF THE MOBILE MODEL

History of the MOBILE Model

The MOBILE model for estimating on-road vehicle emissions factors (in grams per mile [g/mi]) was first developed by the U.S. Environmental Protection Agency (EPA) in the late 1970s. Prior to that time, the agency published simple look-up tables for estimation of mobile-source emissions. The model, originally and still written using the Fortran scientific programming language, has had significant updates and new releases every few years as new data became available, new regulations were promul gated, emissions standards were established, and sources and processes of vehicle emissions were better understood. Each new version of the model has become more complex in the approach to modeling average in-use vehicle emissions, and has provided the user with additional options for tailoring emissions-factor estimates to local conditions. The model versions, release dates, and changes in each model update are summarized in Table 3–1 (EPA 1999e).

Changes in the databases underlying the models and changes in modeling methodology in each successive version result in changes to predicted total on-road vehicle emissions. From one model version to the next, these changes can be either increases or decreases in emissions factors, and the changes are not always in the same direction for all three pollutants (NO_x, CO, and VOCs). Although these changes created somewhat of a moving target for air-quality planners and the regulated industries, the revised models should provide more accurate analyses of mobile source emissions and of the effects of mobile source control programs.

As an example, Figure 3–1 shows emissions for the Baltimore area for calendar years 1988 and 1990 as predicted using three recent official release versions of the model (MOBILE5a, MOBILE4.1, and MOBILE4); these comparisons are unaffected by the inclusion of new emissions standards and regulations. Carbon monoxide (CO) emissions and emissions of nitrogen oxides (NO_x) increase from one model to the next, albeit in different proportions. Volatile organic compound (VOC) emissions, though, decreased from MOBILE4 to MOBILE4.1, and then increased significantly from MOBILE4.1 to MOBILE5a, while still remaining lower than MOBILE4 levels.

MOBILE5—The Current Model

The MOBILE5 model, released in 1993, provides emission factors for on-road vehicles for the three regulated pollutants: VOCs, CO, and NO_x. The model provides emission factors separately for the classes of vehicles listed in Table 3–2, and also for the average on-road fleet using a default national mix of vehicles; the user can optionally input a different fleet mix for the calculation of fleet average emissions. The vehicle classes are further subdivided into technology classes in MOBILE, to account for emissions differences between, for example, vehicles with carburetors and those with fuel injection. To estimate total on-road mobile emissions in a given area, either the vehicle class emissions factor is multiplied by estimates of vehicle miles traveled (VMT) by vehicle class for the area and summed, or the fleet average emissions factor is multiplied by total VMT (across vehicles classes) for the area. These VMT estimates are typically provided by local or regional transportation-planning agencies.

TABLE 3–1 MOBILE Model Revisions

Version	Release Date	Model Revisions
MOBILE1	1978	• Included modeling of exhaust emissions rates as functions of vehicle age/mileage (zero-mile levels and deterioration rates).
MOBILE2	1981	• Updated with substantial data (available for the first time) on emission-controlled vehicles (i.e., catalytic converters, model years 1975 and later) at higher ages/mileages. • Provided additional user control of input options.
MOBILE3	1984	• Updated with substantial new in-use data. • Elimination of California vehicle emissions rates (continued to model low- and high-altitude emissions). • Added tampering (rates and associated emissions impacts) and anti-tampering program benefits. • In-use emissions-factor estimates for nonexhaust emissions adjusted for real-world fuel volatility as measured by Reid vapor pressure (RVP).
MOBILE4	1989	• Updated with new in-use data. • Added running losses as distinct emissions source from gasoline -powered vehicles. • Modeled fuel volatility (RVP) effects on exhaust emissions rates. • Continued expansion of user- controlled options for input data.
MOBILE4.1	1991	• Updated with new in-use data. • Added numerous features allowing user control of more parameters affecting in-use emissions levels, including more inspection/maintenance (I/M) program designs. • Included effects of various new emissions standards and related regulatory changes (e.g., test procedures). • Included impact of oxygenated fuels (e.g., gasohol) on CO emissions.
MOBILE5	1993	• Updated with new in-use data, including basing new basic emissions-rate equations on much larger database derived from state-implemented IM240 test programs. • Included effects of new evaporative emissions test procedure (impact on in-use nonexhaust emissions levels). • Included effects of reformulated gasoline (RFG). • Included effects of new NO_x standard of 4.0 g/bhp-hr for heavy-duty engines. • Included impact of oxygenated fuels on VOC emissions. • Included Tier 1 emissions standards under 1990 Clean Air Act Amendments. • Added July 1 evaluation option.

		Model Revisions
		• Included impact of low-emission vehicle (LEV) programs patterned after California regulations.
		• Revised speed corrections used to model emissions factors over range of traffic speeds.
MOBILE5a	1993	• Corrected a number of minor errors in MOBILE5.
MOBILE5b	1996	• Included final on-board vapor-recovery regulations.
		• Included final reformulated gasoline regulations.
		• Added more user options for I/M programs.

Source: EPA 1999e

All of the MOBILE5 emissions factors are estimated from existing test data, and engineering judgment in the absence of test data. Although there is a detailed User's Guide (EPA 1994) for the model, there is limited documentation from EPA describing the databases and analytical methods used in MOBILE5 to develop the emissions factors.

The user provides inputs (some required and some optional) to MOBILE5 that describe typical operating characteristics, fleet characterization, and mobile-source control programs. These inputs (in addition to vehicle class and VMT mentioned above) are

- ambient temperature;
- average vehicle speeds by vehicle class;
- fuel characteristics; (include fuel volatility and oxygen content, and if reformulated gasoline is in use);
- vehicle inspection and maintenance (I/M) program parameters, if such a program is in place; and
- vehicle age distributions (used to estimate composite emissions across all vehicle model years).

Figure 3–2 shows emissions estimates developed using an updated version of MOBILE5 and updated travel-activity estimates. It shows the predicted distribution of on-road mobile source emissions of VOCs and NO_x in year 2007 by emissions category for New York and Chicago. These inventories were generated by EPA using the latest emissions model developed as part of the regulatory impact assessment for the recent Tier 2 vehicle emissions and fuel sulfur standards (EPA 1999d). This model, known as the Tier 2 Model, was developed from MOBILE 5b and available elements of the upcoming version, MOBILE6 (EPA 1999c). The Tier 2 model is actually a spreadsheet program derived from MOBILE algorithms, outputs, supplemental test data, and assumptions. The MOBILE6 elements incorporated into the Tier 2 Model include updated assessments of in-use vehicle deterioration, fuel sulfur impacts, and fleet characteristics. However, the Tier 2 Model does not use the MOBILE6 methodology and test data for estimating evaporative emissions. MOBILE6 will likely increase the fraction of evaporative to tailpipe emissions of VOCs compared with that obtained in MOBILE5b.

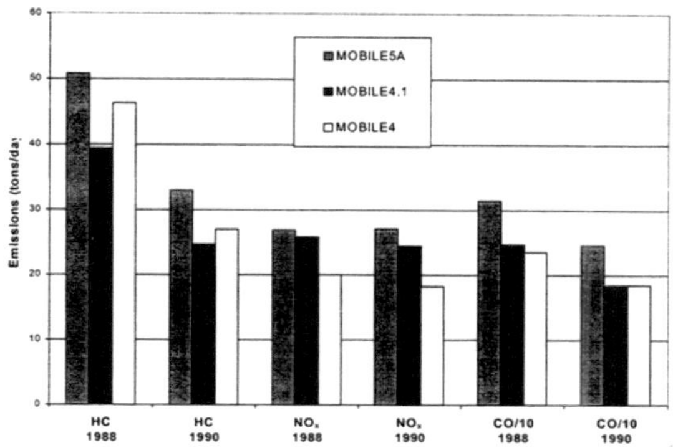

FIGURE 3–1 Comparison of estimated emissions for Baltimore from three recent versions of the MOBILE model. Note that CO emissions are divided by 10. Also note that this report usually uses the term VOCs as opposed to hydrocarbons (HCs) to refer to the general class of gaseous organic compounds

Both cities depicted in Figure 3–2 show a fairly similar emissions profile, although important differences are clear. Chicago has much greater emissions from heavy-duty vehicles. This is especially apparent for NO_x emissions; Chicago has 37% of emissions from HDDVs whereas New York only has 24%. Chicago also has greater emissions from motorcycles. For example, Chicago has 7% of their VOC emissions attributed to motorcycles, over twice the percent of emissions from motorcycles in New York. Generally speaking, the Tier 2 Model estimates that about 45% of the total on-road VOC emissions is from light-duty vehicle exhaust, about 30% is from light-duty evaporative emissions, and the remainder is primarily from heavy-duty vehicles. Note that the regulatory impact analysis was performed for four cities—Atlanta, Charlotte, Chicago, and New York. However, the emissions profiles for Atlanta and Charlotte were similar to those for New York, and are not shown here.

TABLE 3–2 MOBILE5 Vehicle Classes

Vehicle Class	MOBILE Code	Weight Description
Light-duty gasoline vehicles (passenger cars)	LDGV	Up to 6000 lb gross vehicle weight (GVW)
Light-duty gasoline trucks[a] (pick-ups, minivans, passenger vans, and sport-utility vehicles)	LDGT1	Up to 6000 lb GVW
	LDGT2	6001–8500 lb GVW
Heavy-duty gasoline vehicles	HDGV	8501 lb and higher GVW equipped with heavy-duty gasoline engines
Light-duty diesel vehicles (passenger cars)	LDDV	Up to 6000 lb GVW
Light-duty diesel trucks	LDDT	Up to 8500 lb GVW
Heavy-duty diesel vehicles	HDDV	8501 lb and higher GVW
Motorcycles[b]	MC	

[a]Emissions for light-duty trucks are modeled separately for two weight classes with different emissions standards in the Clean Air Act

[b]Highway-certified motorcycles only are included in the model. Off-road motorcycles, such as dirt bikes, are modeled as a non-road mobile source in EPA's NONROAD model.

MOBILE6—The Next Generation Model

EPA's Office of Transportation and Air Quality (OTAQ) has for the last several years been working on the next generation of the MOBILE model, referred to as MOBILE6. This model will be significantly different from MOBILE5 in almost all model components, and will be based on an enormous amount of recent vehicle-emissions testing data from EPA, the California Air Resources Board (CARB), automobile manufacturers, and petroleum refiners. At this point, MOBILE6 is expected to be released in the year 2000. However, EPA has already released substantial documentation and held workshops describing the model revisions, allowing the agency to gather feedback on its proposed modifications. This documentation and public outreach process will be discussed in the following section. The significant changes being incorporated into MOBILE6 include the following:

- dramatically lower basic emissions rates, based on analyses of the Dayton, Ohio I/M program data;
- deparation of start and running-exhaust emissions;
- addition of so-called off-cycle emissions (aggressive driving and air-conditioning operation, which are not included in the Federal Test Procedure [FTP]);

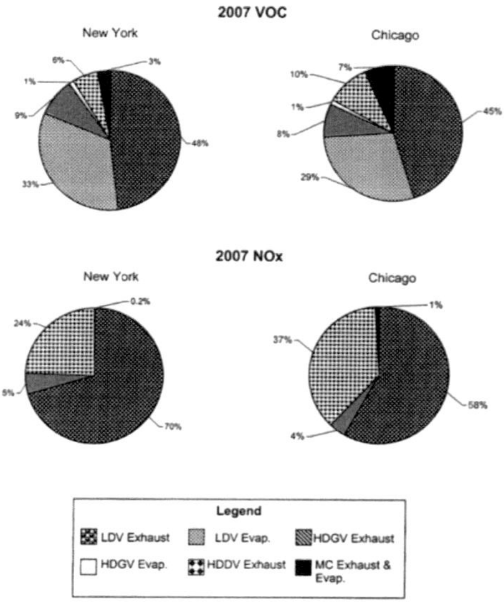

FIGURE 3–2 VOC and NO_x emissions inventories for New York and Chicago. Source: EPA 1999d.

- control of off-cycle emissions with the Supplemental FTP (SFTP) in future years;
- emissions factor estimates for different roadway types (e.g., highways arterials, locals);
- evaporative diurnal emissions factors estimated from real-time diurnal test data previously unavailable;

- revised (lower) estimates of the effects of oxygenated fuels on CO emissions;
- revised (lower) effects of I/M programs on vehicle emissions;
- addition of off-cycle NO_x emissions for heavy-duty diesel vehicles;
- effects of in-use fuel sulfur content on all emissions; and
- effects of national low-emissions vehicle (NLEV) and Tier 2 standards.

Although the MOBILE6 documentation provides numerical results for changes in specific model components, overall changes to average in-use fleet emissions factors will not be known until the full model is released. Thus, it is not yet known whether regional emissions estimates from MOBILE6 will increase or decrease relative to MOBILE5, though it is likely that they will increase at least for VOCs in order to be in better agreement with the findings of evaluation studies that are discussed in Chapter 4.

FEDERAL ADVISORY COMMITTEE ACT PROCESS AND PUBLIC OUTREACH

An important part of the developmental process for MOBILE6 has been public outreach. This includes input from EPA advisory committees, comments from stakeholders and the interested public, and the release of technical documentation describing model modifications.

The Clean Air Act Amendments of 1990 (CAAA90) established the Clean Air Act Advisory Committee to advise EPA on issues of implementation this law. One of the many subcommittees of the Clean Air Act Advisory Committee is the Mobile Sources Technical Review Subcommittee (MSTRS), often referred to as a FACA subcommittee because it is chartered under the Federal Advisory Committee Act (FACA). The MSTRS advises EPA's OTAQ on technical issues specific to the control of emissions from mobile-sources. It is composed of experts on mobile-source emissions from industry, academia, state agencies, and nongovernmental organizations. Meetings are held quarterly and are open to the public.

One of the MSTRS working groups is the Modeling Working Group, which provides on-going advice on the development and improvement of MOBILE and other emissions models. The specific charge for this work group includes helping to set priorities for developments to MOBILE6 and developing procedures for EPA to use when obtaining outside review for products used to support MOBILE6. This group also is producing a comprehensive report on the MOBILE modeling process, problems, and oppor tunities for improvement on which this committee was briefed (this paper, "Big Picture Modeling Issues," is currently in draft form).

Additionally, EPA has increased public feedback and public involvement in the development of MOBILE6 in several other ways. These steps were taken to make the model better understood by the user community and to counter criticism that the model was inadequately documented and peer reviewed. Although the draft MOBILE6 model is not expected to be available until later in 2000, EPA has released detailed technical documentation for most of the proposed changes in the Stakeholder Review Documents on MOBILE6 web page (http://www.epa.gov/OMSWWW/m6.htm). EPA is to be commended for documenting the databases used and the development of revised emissions factors in the Stakeholder Review Documents. This documentation, although not always complete in describing the full details of the analyses, is a major improvement from all previous versions of MOBILE. EPA provides a 60-day review period for each document as it is posted, and has stated its intention to provide all comments and responses to comments with each document in the final version.

EPA has also held three workshops discussing the new version of the model and created an e-mail list server to update interested parties on new model developments. The workshops were open to the public. They were intended to update interested parties on EPA's plans for the model as well as solicit input and reaction to those plans. The workshops included both technical presentations describing changes to the model methodology and presentations oriented to model users describing changes to data input and output. The e-mail list server is used to announce the workshop agendas, the release of new documentation concerning the MOBILE6 model, updates to the current version of MOBILE5b, and other information.

RELATED MODELS

There are several emissions models and databases related to EPA's MOBILE model, which are used to estimate mobile-source emissions inventories and provide inputs for air-quality models. These are

- **PART5**—estimates particulate matter (PM) emissions factors for on-road vehicles;
- **Complex Model**—estimates emissions impacts of reformulated fuel compositional changes on 1990 light-duty gasoline vehicles;
- **MOBTOX**—estimates on-road mobile-source toxic emissions factors; and
- **SPECIATE, and related databases and models**—provide VOC speciation profiles for complex photochemical grid modeling.

These models are each described briefly below. References are provided for readers who desire more detailed information on any of these models.

PART5

EPA's PART5 model estimates PM emissions factors in g/mi for 12 vehicle classes. Emissions factor estimates are provided for particle diameter sizes from less than or equal to 1.0 to 10.0 μm (micrometers). The model is referred to as PART5 to indicate consistency with MOBILE5 in fleet characterization data and in the general methods used to estimate basic emissions rates. PART5 estimates all PM emissions associated with on-road travel: exhaust emissions, brake-wear emissions, tire-wear emissions, and fleet-average paved and unpaved road-dust emissions. For HDDVs, PART5 also provides estimates of idle emissions. The PART5 model has been updated in only very minor ways since the original model development in the mid-1980s. The emissions factors in the model are based on either engine certification data or on ratios from VOC emissions. Although there is a User's Guide for the model, there is no documentation that explains the derivation of the emissions factors in the model.

The emissions-factor estimates in the PART5 model are seriously out of date. New PM test data have recently become available for both light-duty vehicles (LDVs) and heavy-duty vehicles (HDVs). Although on-road mobile-source emissions-factor estimates will be needed for state and local air-quality planning agencies to develop PM emissions inventories and air-quality management plans for the new PM standards, EPA's OTAQ has not focused as much effort on updates to the PART5 model as on MOBILE. However, many of the revisions developed for MOBILE6 can be easily incorporated into PART5, and OTAQ has done so for estimating PM emissions as part of the Tier 2 rulemaking (EPA 1998d). But the major improvement required for the model is the inclusion of recent testing data for the revision of emissions factors.

Because most of the methods for estimating PM emissions in PART5 are similar to the methods used in MOBILE5, and because PART5 needs major revision, an updated version of MOBILE6 should incorporate revised PM emissions-factor estimates. For most users, it would be much more desirable to have one integrated model that provides emissions-factor estimates for PM as well as VOC, NO_x, and CO.

Complex Model

The Complex Model is used by petroleum refiners and other interested parties to estimate how gasoline composition affects vehicle emissions.

The model was developed in a regulatory negotiation process between EPA and the affected industry. The model is fully described in EPA's reformulated gasoline Regulatory Impact Analysis dated December 13, 1993 (EPA 1993b). The Complex Model, which is a spreadsheet model downloadable from EPA's reformulated gasoline web page (http://www.epa.gov/OMSWWW/rfg.htm), predicts percent change in 1990 technology vehicle emissions for a target reformulated gasoline (RFG) relative to U.S. 1990 baseline gasoline. Emissions are a function of the following input parameters:

- MTBE (methyl tertiary-butyl ether, weight percent oxygen [wt%]),
- ETBE (ethyl tertiary-butyl ether, wt%),
- ethanol (wt%),
- TAME (tertiary-amyl methyl ether, wt%),
- sulfur (parts per million [ppm]),
- RVP (Reid vapor pressure, pounds per square inch [psi]),
- E200 (percent of fuel that evaporates at 200° F),
- E300 (percent of fuel that evaporates at 300° F),
- aromatics (percent by volume),
- olefins (percent by volume), and
- benzene (percent by volume).

U.S. baseline emissions are calculated from MOBILE5 runs with U.S. industry-average gasoline. The model calculates changes from baseline emissions to emissions for the target fuel for exhaust and evaporative VOCs, air toxics (benzene, formaldehyde, acetaldehyde, and 1,3-butadiene), and exhaust NO_x. (The model does not estimate the effects of fuel reformulation on exhaust CO emissions.) The model is a statistical model based on testing data from several major programs measuring the emissions effects of the various fuels tested.

The present and planned versions of MOBILE allow for the specification of limited fuel properties (e.g., whether or not RFG is used) but do not allow for the specification of the detailed fuel properties that are available in the Complex Model. This means that if states or nonattainment areas choose to require a fuel with greater emissions reductions than required of federal RFG, the Complex Model must be run first to generate scaling factors to apply to MOBILE output. Even then, there are questions as to how the results would be used given that COMPLEX was developed for 1990 vehicle technologies only. The fuel effects now calculated in the Complex Model should be updated so that they can appy to all model years and technology groups, not just 1990 technology vehicles, and added to future versions of MOBILE. CARB's Predictive Model estimates fuel effects for all on-road light-duty vehicles based on a broader database than that used in the development of the Complex Model; EPA should review and consider updating CARB's model. With this addition, local variations in gasoline properties can be more easily modeled for use in air-quality planning.

It is important to also include CO emissions within the Complex Model. Historically, the RFG program was used for ozone reduction and the Complex Model was used as a tool for certifying the performance of RFGs with respect to VOCs, air toxics, and NO_x. CO was not a component of the Complex Model because CO was not required for evaluating an RFG. However, there are some CO nonattainment areas that are developing RFGs. Additionally, CO is an ozone precursor.

MOBTOX

The MOBTOX model is used for estimating toxic-emissions factors for on-road motor vehicles. The model was originally developed as part of EPA's CAAA90 mandated study on "motor vehicle-related toxics", and is now being updated for use in both the regulatory impact analysis for the Tier 2 vehicle standards rulemaking and the development of regulations for controlling mobile-source toxic emissions (EPA 1999f). The first version of the model was based on MOBILE4.1. The next version, called MOBTOX5b, was based on the modified version of MOBILE5b used in EPA's July 1998 Tier 2 study (EPA 1998d). The modified MOBILE5b model includes alternative basic emissions rates, and the effects of aggressive driving and air-conditioning usage.

MOBTOX5b applies exhaust and evaporative toxic-adjustment factors for various vehicle classes and technologies to MOBILE5b VOC emissions factors. It estimates emissions factors for benzene, formaldehyde, acetaldehyde, 1,3-butadiene, and methyl tertiary-butyl ether (MTBE). For benzene and MTBE, the model estimates factors separately for exhaust, evaporative (diurnal and hot-soak), refueling, running-loss, and resting-loss emissions. The model also has the capability to account for differences in exhaust VOC toxic fractions between normal- and high-emitting vehicles. The toxic-adjustment factors for newer technology vehicles were developed from the speciation data for 1990 technology LDVs developed for the Complex Model.

The MOBILE revisions in MOBTOX5b, developed before many of the MOBILE6 revisions were proposed, are different from what is now proposed for MOBILE6. Although the version of MOBTOX5b first developed for the Tier 2 rulemaking is publicly available, EPA is currently updating the air toxics model to more closely reflect MOBILE6 proposals. The updated model will be publicly released with the final Tier 2 rule.

As with the PART5 model, MOBTOX must use the same vehicle-activity data as MOBILE, but has different emissions factors for toxic species. For both user convenience and model consistency, these toxic-emissions factors should be incorporated into a future version of MOBILE and the use of the separate MOBTOX model should be discontinued.

NONROAD

EPA is currently developing a national emissions model for off-road mobile-sources. The current draft version of the model, called NONROAD, is available on EPA OTAQ web page (http://www.epa.gov/oms/nonrdmdl.htm). The web page contains full documentation for the model, including a User's Guide and detailed technical documentation for all model estimates, inputs, and assumptions.

Unlike the MOBILE model, NONROAD provides activity data as well as emissions factors. Thus, it can specifically estimate emissions inventories for off-road equipment. The model predicts emissions for vehicles and equipment types in the following categories:

- airport ground support, such as terminal tractors;
- agricultural equipment, such as tractors, combines, and balers;
- construction equipment, such as graders and back hoes;
- industrial and commercial equipment, such as fork lifts and sweepers;
- recreational vehicles, such as all-terrain vehicles and off-road motorcycles;
- residential and commercial lawn and garden equipment, such as lawn mowers and leaf and snow blowers;
- logging equipment, such as shredders and large chain saws;
- recreational marine vessels, such as power boats;
- underground mining equipment; and
- oil field equipment.

The model includes more than 80 basic and 260 specific types of nonroad equipment, and further stratifies equipment types by horsepower rating. Fuel types include diesel, gasoline, compressed natural gas, and liquefied petroleum gas. The model estimates six exhaust emissions (VOC, NO_x, CO, carbon dioxide (CO_2), sulfur dioxide (SO_x), and PM), and also estimates nonexhaust VOC emissions for six modes—hot-soak, diurnal, refueling, resting-loss, running-loss, and crankcase emissions. The model can estimate national total emissions, emissions by state, or for one or more counties or subcounties in one or more states. A more complete description of the NONROAD model's capabilities can be found in Pollack and Lindhjem (1998) and in the NONROAD User's Guide (ENVIRON 1998).

The NONROAD model currently does not include emissions estimates from locomotives, aircraft, and commercial marine engines, and these will not be included in the official release of NONROAD. EPA is currently gathering the data necessary to estimate emissions for these important sources and plans initially to provide written guidance to states on developing emissions inventory estimates for these sources based on these data. At present, EPA expects to have draft guidance documents available in 2000, with the final guidance documents available later that year. Release of NONROAD software modules for these applications is not expected before 2001.

EPA is currently addressing comments on the NONROAD draft model, and has not yet had the model peer-reviewed. The current draft is therefore likely to undergo revision before the final version is released, which is expected to be sometime in 2000. As future Tier 2 vehicle standards and corresponding sulfur-reduction regulations reduce on-road mobile-source emissions, non-road emissions will become a larger fraction of the total emissions. The NONROAD model is extremely data driven, and there are many gaps in the available data. EPA should place more emphasis on improving both the emissions factors and activity data in this model.

SPECIATE and Emissions Processing Systems

Photochemical air-quality models require emissions of specific types of VOCs (e.g., formaldehyde, acetaldehyde, alkenes, aromatics, short-chained alkanes, and higher alkanes), as opposed to the total VOC mass provided by MOBILE. Providing such speciation information is done in two steps. First, the total VOC emissions, which are estimated using MOBILE and the activity data, are split into about 100 individual organic compounds and their emissions rates are determined (e.g., formaldehyde, ethane, and propane). Next, these compounds are grouped into a smaller number of lumped organic species used by air-quality models. (Using all of the individual compounds would be excessively burdensome in most photochemical modeling applications, although it is possible.)

Speciation is accomplished by using tables of profiles that have been developed from source testing. One such set of profiles for VOCs and PM has been developed by EPA, and is part of the SPECIATE system. The database and User's Guide are on EPA's web page at http://www.epa.gov/ttn/chief/software.htm/#speciate. This database is now extremely out of date, especially for mobile-source emissions. On the web page, EPA itself has indicated serious concerns with the database. Recently, a number of emissions processing systems for producing spatially and temporally allocated speciated emissions rates for all emissions sources have been developed by contractors and universities. These include the Emissions Processing System (EPS2) and the Emissions Model ing System (EMS95). The systems have a library of profiles, some of them originally based on SPECIATE, but with enhancements and additions.

For specific air-quality modeling applications, users enhance the available speciation profile databases and models for their own use. For example, the Auto/Oil Air Quality Improvement Research Program conducted a large test program to develop speciated automobile emissions rates, and used that information in follow-on air-quality modeling studies. CARB, using those measurements and others, has developed their own set of emissions profiles as well.

PREVIOUS REVIEWS OF MOBILE

Several technical reviews have been done for different versions of the MOBILE model. The reviews were inspired by field observations that indicated a disagreement between model predictions and actual emissions measurements. Industry groups and agencies outside EPA have sponsored these reviews, primarily because the databases and methods underlying the models have not been well documented by EPA. The reviews included examination of the model's structure, assumptions, sensitivity to changes in model parameters, method of accounting for I/M emissions reductions, and the effects of model revisions on emissions inventories.

Summary of Reviews

The first review was performed for the Coordinating Research Council (CRC, which sponsors research for the automobile and oil industries) by Pollack et al. (1991). This review was prompted by studies of field measurements (in particular, tunnel studies performed as part of the Southern California Air Quality Study) that indicated that the mobile-source emissions-factor models developed by EPA and CARB substantially underpredicted emissions levels (Ingalls et al. 1989). To assist in understanding the potential sources of model underprediction, CRC sponsored a project to evaluate and compare the data and methodologies used in the then-current versions of the emissions-factor models—EPA's MOBILE4 and CARB's EMFAC7E. A detailed comparative review was conducted of the databases and methods used to derive exhaust and evaporative emissions. The project report evaluated relative strengths and weaknesses of the models, and discussed potential causes of emissions underprediction. In addition, sensitivity analyses were performed to understand changes in model predictions in response to changes in model inputs and fixed model parameters.

The American Petroleum Institute (API) sponsored reviews of

MOBILE5a aimed at evaluating the basis and validity of updates from the previous version, with a focus on assumptions and extrapolations in the model. Heiken et al. (1994) critically reviewed the model's basic exhaust emissions rates for LDVs and HDVs, and all nonexhaust emissions rates (hot-soak, diurnal, running-loss, resting-loss, refueling). Vehicle test data, source code, and EPA internal memoranda were reviewed to replicate the methodology used to develop the model's algorithms and input data. Statistical approaches for developing equations were also reviewed. In response to API interests, the report included a detailed review of the effects of fuel oxygenates on exhaust and evaporative emissions. Sensitivity analyses were performed to evaluate the effects of alternate assumptions and updated methodologies used for MOBILE5a.

API also funded Sierra Research to review specific components of MOBILE5a. The first study evaluated and documented MOBILE5a methods for estimating the benefits of I/M programs and adoption of the California Low Emission Vehicle (LEV) standards (Sierra Research 1994a). The evaluation included an analysis of the data that EPA used to develop I/M identification and repair effectiveness, and an assessment of the methodology used to estimate the benefits of evaporative system functional checks. The primary conclusion of the report was that the model likely overestimated the effects of enhanced I/M programs for both exhaust and evaporative emissions. The second Sierra Research study was focused on an important update in the MOBILE5a model-the use of I/M test data to develop LDV basic exhaust emissions rates (all previous versions of the model were based on EPA's FTP testing). Because most of the exhaust emissions-correction factors (i.e., temperature, speed) were based on the FTP, EPA developed a conversion from the IM240 test data to FTP for calculating the basic emissions rates. The Sierra report critically reviewed the IM240-to-FTP conversion process, and checked the procedure with a second IM240 data set from Mesa, Arizona (Sierra Research 1994b).

The U.S. Department of Transportation's (DOT) Federal Highway Administration (FHWA) also sponsored a review and evaluation of different versions of the MOBILE model. DOT's interest was prompted by requirements in the CAAA90 and the 1991 Intermodal Surface Transportation Efficiency Act (ISTEA) for local transportation-planning and conformity determinations. Because MOBILE was the modeling tool to be used to estimate transportation emissions, DOT desired an understanding of the structure and operation of MOBILE, and a documentation of the changes that had occurred among model revisions. The report prepared for FHWA (Sierra Research 1994c) included an explanation of the basic parameters of the MOBILE model for individuals with little background in motor-vehicle emissions modeling. Detailed descriptions were then provided on the calculation of exhaust and evaporative emissions, methods used to determine I/M program effectiveness and other CAAA90 requirements (e.g., Tier 1 vehicle standards and reformulated and oxygenated gasoline). The report also included an evaluation of the changes in fleet average emissions factors as predicted by MOBILE4, MOBILE4.1, and MOBILE5a.

General Accounting Office Review of MOBILE5

The genesis of the 1997 report prepared by the General Accounting Office (GAO) on the MOBILE model (GAO 1997, hereafter called the GAO report) was mentioned briefly in Chapter 1 of this report. The GAO report raised 14 specific concerns about the model. During the preparation of that report, EPA provided a response to an initial draft and noted the issues that they planned to address in the MOBILE6 update. Table 3–3 lists the issues raised about the MOBILE model in the GAO report and shows that EPA planned to address all of those issues except for road grade, heavy-duty I/M, and model uncertainty in the development of MOBILE6. In the following sections of this chapter, the GAO report issues listed in Table 3–3 are discussed and the improvements planned in MOBILE6 to address the issues are evaluated.

HIGH EMITTERS

Underrepresentation of emissions from high emitters in MOBILE, even in the emerging MOBILE6, is considered to be one of the chief reasons for MOBILE underpredicting real-world fleet emissions. In general, these vehicles are difficult to characterize statistically because the number of high emitters is relatively small and the range in their emissions is relatively large. Priority should be given to further improving this very important component of MOBILE.

Exhaust High Emitters

The problem of high emitters, and their correct representation in the databases for MOBILE, was identified as Issue 6 in the GAO report. Characterizing high emitters requires understanding of not only their level of emissions but also their population and activity. High emitters will generally represent a disproportionately high fraction of the total fleet emissions predicted by MOBILE. Concerns have been expressed that, because of potential recruitment bias in FTP testing used as the foundation of MOBILE, the model underestimates emissions of the overall real-world fleet. Typically, recruitment acceptance rates are less than 25%. It is theorized that owners of vehicles that are high emitters will be reluctant to submit their vehicles for intense emissions testing. Conversely, domestic automobile manufacturers, who have supplied much of the later model-year FTP data for use in developing MOBILE6, have expressed the opinion that their recruitment program, which offered free repairs, would actually

TABLE 3–3 Major Limitations in MOBILE5a Model Identified in GAO Report

Areas of concern regarding MOBILE cited in GAO report		Does EPA plan to address this issue in MOBILE6?
1.	Emissions estimates for higher speeds, especially speeds in excess of 65 miles per hour (mph).	Yes
2.	Representation of emissions from rapid acceleration and deceleration, including aggressive driving behaviors.	Yes
3.	Representation of emissions immediately after engine start-up, known as cold-start emissions.	Yes
4.	Representation of emissions from air conditioner use.	Yes
5.	Representation of emissions from road grades, such as when a car climbs a hill.	No
6.	Representation of high-emitting vehicles in MOBILE's supporting database.	Yes
7.	Representation of emissions from lower-polluting fuels, especially fuels with lower volatility.	Yes
8.	Representation of emissions-system deterioration for vehicles with 50,000 or more odometer miles.	Yes
9.	Emissions estimates and assumptions for vehicle I/M programs.	Yes
10.	Estimates and assumptions for nontailpipe evaporative emissions when the vehicle is not operating.	Yes
11.	Emissions estimates and assumptions for the I/M of HDVs—those with a gross vehicle weight of 8,501 pounds or more.	No
12.	Data characterizing vehicle fleet.	Yes
13.	Greater distinctions in roadway classifications.	Yes
14.	Quantifying the uncertainty of the model's estimates.	No

encourage recruitment of higher emitters. Even if the latter is true, the data submitted by domestic automobile manufacturers are composed of vehicles roughly 2–3 years old where high emitting vehicles should be a very small fraction of these vehicles. The California Air Resources Board has used a high-emitter correction factor in its mobile-emissions model (EMFAC) for a number of years. This correction factor is based on random roadside emissions testing surveys of the real-world fleet and has historically resulted in increasing the emissions estimates of EMFAC.

Source: GAO 1997.

EPA decided to use a high-emitter correction factor in development of MOBILE6. EPA selected first-year IM240 test data from Ohio for development of this correction factor (EPA 1999g). The advantage of this approach is that it presents a relatively unbiased sample of the real-world fleet. EPA did recognize the disadvantages of using IM240 data compared to FTP testing, which are described in Chapter 4. EPA defined high emitters as those vehicles emitting more than twice the FTP standards for VOCs and NO_x and more than three times the FTP standard for CO. Vehicles in these emissions categories were felt to have significant problems with their emissions control systems. The high-emitter correction factors developed by EPA for MOBILE6 for running emissions result in a significant increase in projected emissions, ranging from +30% for CO to +150% for VOC for 1988–1993 port fuel-injection cars. EPA also used the high-emitter correction factor developed for running emissions to adjust the start-emissions component of MOBILE6 (EPA 1999h) and also to address FTP recruitment bias.

EPA's use of IM240 data to generate a high-emitter correction factor for MOBILE6 appears to be a step in the right direction toward improving the accuracy of MOBILE. Justifying this approach is the fact that numerous tests of MOBILE's validity tend to indicate MOBILE has been underestimating real-world fleet emissions (see discussions in Chapter 4), and EPA's new high-emitter correction factor result in increasing the base emissions rates in MOBILE6.

As discussed in the evaluation section of Chapter 4, there are many reasons to suspect that even IM240 data fall short of truly reflecting the real-world fleet emissions, particularly because it is not truly representing the high-emitter category. For example, remote-sensing data, discussed in Chapter 4, indicates the shortcomings of IM240 data representation of high emitters. Forthcoming roadside-pullover loaded-mode testing being conducted by the California Bureau of Automotive Repair might provide the greatest insight into how well MOBILE6 reflects the contribution of high emitters to total fleet emissions.

Because start emissions are a major component of the total fleet emissions projected by MOBILE, the accurate characterization of start emissions from high emitters is critical. EPA has acknowledged that confidence intervals around the average start-emissions levels of the high emitters are quite large due to high scatter and small sample sizes. Additionally, the representativeness of applying the IM240 high-emitter correction factors developed for running emissions to start emissions must be questioned. Although using the running-emissions correction factor to correct start emissions is undoubtedly better than applying no correction factor, running-emissions correction factors would be less likely to reflect the increased catalyst light-off temperatures associated with older, higher-emitting vehicles. Higher catalyst light-off temperatures, one of the common characteristics of high emitters, would cause the adjustment for cold-start emissions to be relatively higher than the adjustment for running emissions. This is because running emissions are measured for a fully warmed engine.

Evaporative High Emitters

As is the case for tailpipe emissions, the distribution of evaporative emissions among the in-use vehicle fleet is highly skewed towards high emitters. Skewness is characteristic of all three evaporative emissions types: hot-soak, running-loss, and diurnal. Because the measurement of evaporative emissions is difficult, available data are limited and come primarily from a series of studies sponsored by the CRC. EPA has used these data (as well as data obtained by EPA) in the formulation of MOBILE6 that has a special treatment for high evaporative emitters (EPA 1999i). These relatively recent evaporative emissions measurements suggest that evaporative emissions of VOCs are greater than tailpipe emissions. Although older model, carbureted vehicles typically have higher evaporative emissions than newer model, fuel-injected vehicles, high emitters are found in the newer as well as older model vehicles. Running-loss emissions resulting from liquid leaks are probably the most important cause of evaporative high emitters.

DRIVING-CYCLE ISSUES

Driving Cycles—Real-World Driving versus FTP Speeds, Accelerations, and Other Engine-Load Conditions

The FTP for LDVs was designed as a certification test for measuring the emissions from new vehicles. As such, it is a compromise between the desire to have a representative sample of actual vehicle operating conditions and the requirements of actual testing. The latter requirements include equipment limitations, test costs, and test time.

The need to measure vehicle emissions over a range of conditions that simulate actual driving has been recognized since the start of vehicle emissions-control programs. However, the original certification test cycle used only steady-state operation and did not include start emissions. The data for the present FTP driving cycle was obtained from driving survey data taken in downtown Los Angeles during the 1960s. The basic test cycle, which was first used in 1972, measures transient vehicle emissions, including a cold start. A hot-start portion was added in 1975.

The test cycle simulates driving a route of 7.5 miles with an average speed of 19.6 miles per hour (mph). The weighted sum of a cold-start (12-hr engine-off time or soak time at 75°F) and a hot-start (10-min soak time) trip over this route is used to compute the overall emissions. In practice, the trip is divided into two parts. The first part represents transient emissions after start. The test for this part lasts for 505 seconds and covers 3.59 miles. The second part represents stabilized emissions with a warmed engine and catalyst. This part lasts 867 seconds and covers 3.91 miles. The average speeds of the transient and stabilized parts are 25.6 mph and 16.2 mph.

In the actual measurement, the stabilized part is measured only one time, immediately following the cold-start test, and the results of this measurement are assumed to apply to both the hot-start and the cold-start trips. The measured emissions consist of three parts: Bag 1, representing the cold-start transient emissions; Bag 2, representing the stabilized emissions after the engine is warm; and Bag 3, representing the hot-start transient emissions. The certification emissions from the FTP are computed in the following manner:

Average emissions=43% (cold-start trip)+57% (hot-start trip)
Cold-start trip (g/mi)=[(3.59 mi) (cold-start phase g/mi)+ (3.91 mi) (stabilized phase g/mi)]/(7.5 mi)
Hot-start trip (g/mi)=[(3.59 mi) (hot-start phase g/mi)+ (3.91 mi) (stabilized phase g/mi)]/(7.5 mi)

Combining these three equations gives the final weights for each phase in the certification test procedure.

Average emissions=0.206 (cold-start phase)+0.521 (stabilized phase) +0.273 (hot-start phase) (3–1)

The certification results, with units of grams per mile, do not provide any spatial location for start emissions. The separation of start and running emissions proposed for MOBILE6 is discussed later.

At the time the cycle was established, limitations of existing dynamometers restricted the range of possible accelerations. Thus the maximum acceleration in the FTP was limited to 3.3 mph/s. The basic speed-time trace for the 7.5-mile trip is shown in Figure 3–3. This cycle is sometimes referred to as the "LA4" cycle or the urban dynamometer driving schedule (UDDS).

The selection of a downtown route limited the speeds in the FTP. As Figure 3–3 shows, there is a small portion of the cycle in which vehicle speeds exceed 50 mph. Otherwise, the preponderance of the vehicle speeds is below 30 mph. In addition, the limitation of the acceleration rate to 3.3 mph/s does not provide measurements over higher accelerations that are experienced in everyday driving. This implies that the FTP is not representative of modern urban driving.

These limitations in the demand placed on the vehicle and its engine in the test cycle are important because emissions increase with engine load. The emissions of a particular pollutant from engine operation are equal to the product of the mass flow rate of the exhaust and the mass fraction of the pollutant species. As the load on the engine increases, more fuel and air are required, producing a higher mass flow rate of exhaust. In addition, when optimizing power at high engine loads, the fuel metering system in a typical vehicle will supply the engine with a fuel-rich mixture causing a significant increase in the mass fraction of VOC species and CO. Thus higher engine loads—especially those above the loads tested in the FTP—can lead to very large emissions rates.

The GAO report on the MOBILE model identified three issues related to the underrepresentation of high-load conditions in the model. The first two of these were the absence of high-speed driving conditions and the lack of "aggressive" driving—operating conditions with high amounts of acceleration and deceleration. Revisions to the MOBILE model should address these considerations as discussed below. A third issue, the absence of any accounting for road-grade effects, is discussed in a separate subsection below. These effects can be better understood in terms of the equations for engine load presented below.

The load on the engine of a vehicle can be expressed in terms of its various components by the following equation (Bauer 1996):

$$p = \frac{1}{\eta_{dt}}\left[p_{rf} + p_{ad} + p_a + p_g\right] + p_{acc}$$

(3–2)

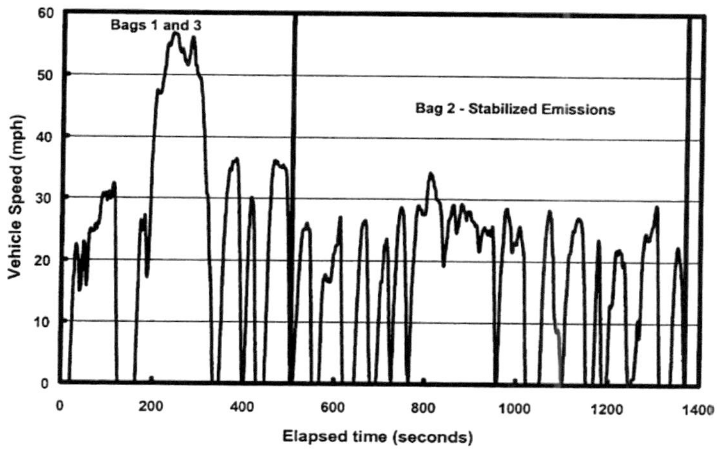

FIGURE 3–3 FTP driving cycle for light-duty vehicles. Source: Sierra Research 1994b.

where
p=the engine power requirement,
η_{dt}=efficiency of the drivetrain,
p_{rf}=power required to overcome rolling friction resistance,
p_{ad}=power required to overcome aerodynamic drag,
p_a=power required to accelerate the vehicle,
p_g=power required for grade climbing, and
p_{acc}=power required by vehicle accessories (such as air-conditioning).
These individual power components are given by the following equations:

$$P_{rf} = f W V ; P_{ad} = \frac{\rho_a C_D A_f V^3}{2} ; P_a = \frac{a}{g} W V ; P_g = W V \frac{b}{\sqrt{1 + b^2}}$$

(3–3)

where
f=coefficient of rolling friction,
W=weight of the vehicle,
V=vehicle speed,
ρ_a=density of air,
C_D=aerodynamic drag coefficient for the vehicle,
A_f=frontal area of the vehicle,
α=vehicle acceleration,
g=acceleration of gravity, and
b=slope of the grade.

These equations show that the power requirements for a given vehicle (and hence the emissions in grams per second) increase with increases in the speed, acceleration, and grade. In addition, the role of grade is seen to be similar to that of acceleration; the power required to climb a 10% grade is essentially the same as the power required to accelerate at 10% of the acceleration of gravity. The most significant accessory load is that of the vehicle-air conditioner.

The increased exhaust flow with increased load is only one source of increased emissions. In addition, there is the potential that under such operating conditions the computer emissions-control systems will perform less effectively than during operation within the FTP speeds and loads. This raises two significant issues: (1) are vehicle emission controls as effective as implied by the FTP-based emission standards, and (2) how realistic are the MOBILE emission estimates based on FTP data?

The technical community has long known about the absence of high speeds and accelerations from the FTP. The CAAA90 required EPA to develop a new emissions test that accounted for this real-world driving. In response to this mandate, EPA sponsored various studies to determine the pattern of vehicle speeds and accelerations that are encountered in everyday driving. Based on these studies, they have developed a Supplemental Federal Test Procedure (SFTP). This SFTP procedure requires vehicle manufacturers to certify vehicles over an additional test cycle known as the US06 cycle. This cycle, whose speed-time trace is shown in Figure 3–4, has speeds as high as 80 mph and a maximum acceleration of 8.4 mph/s. The average speed of this cycle is 48.3 mph. Certification to this new cycle will be phased in between the 2000 and 2004 model years. This test procedure should ensure that vehicle emissions control systems will provide improved emissions control over a wider range of vehicle speeds and loads. Much of the improved emission control will come from reduced use of fuel-rich mixtures at higher loads. Although this cycle does not address the issue of simulating real-world driving in MOBILE, it does include some observed driving speeds and accelerations that are much higher than those used in the FTP.

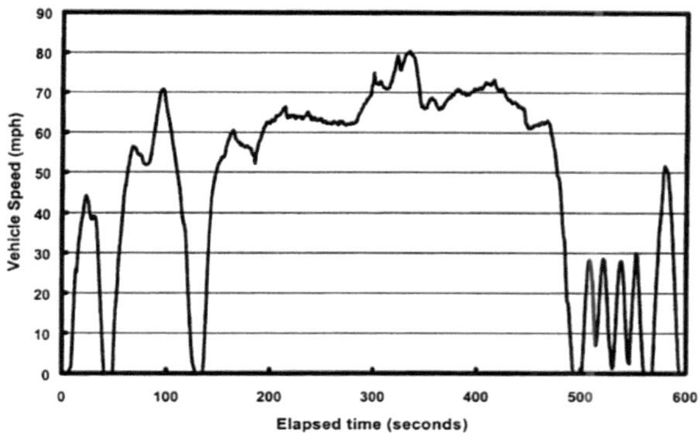

FIGURE 3–4 The EPA supplemental driving schedule (US06). Source: Sierra Research 1994b.

Accounting for Road Grade

Equations 3–2 and 3–3 show that road grade plays a role that is as important as vehicle acceleration. The effects of road grade are not included in the FTP or the US06 cycle. It is not a part of the MOBILE model, and EPA does not plan to add road grade effects to MOBILE6. From a stand-point of vehicle certification, the high-acceleration loads in the US06 cycle should ensure that vehicle emissions-control systems would be operative during grade-climbing operations in real-world driving. However, MOBILE6 will not be able to model the effects of road grade on emissions in a local area. This will be particularly important in urban areas, such as

Denver and Spokane, which have a significant amount of vehicle operation at grade.

In the development of the US06 cycle, EPA obtained some information on driving conditions at grade, but these were not incorporated in the final certification cycle. One possible approach for including grade operations in a future version of the MOBILE model would be the use of grade-correction factors, similar to the facility-correction factors discussed below. However, the use of such factors would require the local planning agencies to specify the amount and type of driving on grades. EPA has stated that they plan to include grade operation in a future version of the MOBILE model. To make such a future model useful to local planning agencies, EPA should discuss with them the types of data that they require on grades so that the model is designed to best use available data. The importance of grade can be estimated from modal or instantaneous models currently under development (Guenstar et al. 1998; Barth et al. 1998)

Speed-Correction Factors and Facility Driving Cycles

The present version, MOBILE5, as well as some earlier versions have used speed-correction factors (SCFs) to estimate emissions under operating conditions that are different from the FTP. These SCFs are based on a series of test cycles with different mean speeds. The FTP emissions (at the mean FTP speed of 19.6 mph) are multiplied by the SCF for a desired speed to give the emissions at the desired speed. SCF are a function of vehicle type, model year, and pollutant species.

SCFs are applied to emissions rates expressed in terms of grams per mile. This emissions rate per unit distance, M_D, is simply related to the mass flow rate of the pollutant per unit time, $\dot{M}$, and the vehicle speed, *V*, by the following equation:

$$M_D = \frac{\dot{M}}{V}$$

(3–4)

Although increased engine loads will cause the emissions rate, $\dot{M}$, to increase with speed, the gram-per-mile emissions rate might increase or decrease depending on the relative changes in mass emissions rate and vehicle speed. This effect can be seen in the SCFs shown in Figure 3–5 (Sierra Research 1994c.) This figure shows the SCFs for 1990 model-year gasoline-powered passenger cars. In this figure, SCFs for all species is 1 at the mean FTP speed of 19.6 mph. For low speeds the SCFs increase dramatically. At these low speeds, the emissions rate is approaching the idle mass emissions rate and the division by a small velocity causes a large emissions rate in grams per mile. Note that expressing emissions factors in units of g/mi creates a problem in describing vehicles at rest (idling).

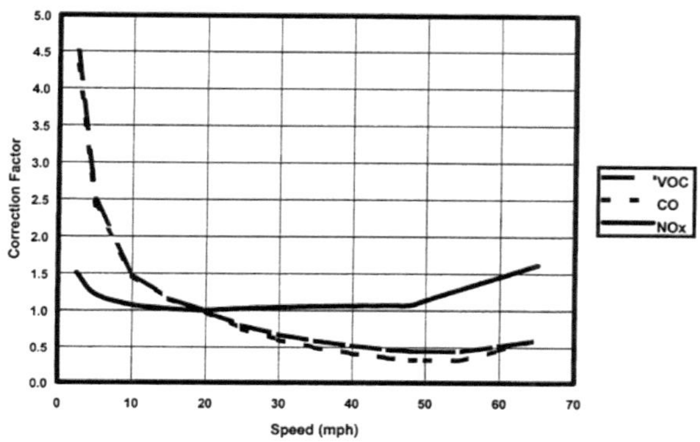

FIGURE 3–5 MOBILE5 speed-correction factors for 1990 model-year, gasoline-powered passenger cars. Source: Sierra Research 1994c.

For vehicle speeds between 20 and 50 mph (for VOCs and CO) or 0 to 20 mph (for NO_x), the SCF decreases, showing a decrease in gram-per-mile emissions with speed. In this region, the mass emissions rate is actually increasing slightly, but not as fast as the speed. At high speeds, the increase in emissions rate is larger than the increase in speed, so the SCF increases.

The driving cycles for the various intermediate speeds that are used to determine the SCFs assume that all types of driving can be characterized by a single parameter, the vehicle speed. However, actual emissions also depend on the engine load, which is determined not only by the vehicle speed, but by a combination of vehicle speed and acceleration. The SCF approach, which assumes that all variation in operating conditions can be characterized by a single parameter, the vehicle speed, would be appropriate for emissions inventories if the speed correction factors in fact ac counted for all types of vehicle operation (see the description in Chapter 5 of an approach that relates emissions to vehicle-specific power). However, this approach is not useful for roadway modeling where the engine operation will be significantly different, for a given vehicle speed, depending on the roadway type. For example, an average speed of 30 mph on a surface street would indicate free-flowing traffic. However, the same average speed of 30 mph on a freeway could indicate congested driving with significant amounts of acceleration and deceleration and significantly higher emissions.

EPA will be using facility-correction factors in MOBILE6 (EPA 1999j). These correction factors will be able to account for differences in overall vehicle operation, on a number of roadway types, under different levels of congestion. Sierra Research has developed specific cycles for these different facility types based on real-world driving data (Austin and Carlson 1997). EPA has used these cycles to obtain data on the difference in emissions between FTP operations and the operations on the facility cycles. Various parameters, including maximum speed and acceleration for these cycles, are shown in Table 3–4.

The application of facility-correction factors in MOBILE6 is similar to the use of SCFs. Instead of having a single set of SCFs, MOBILE6 will have factors that account for both speed and facility type. EPA has used emissions measurements on the various cycles to develop the ratio of emissions on a particular facility type to emissions on the FTP. Once the FTP emissions are computed by the usual MOBILE equations, the emissions for a particular facility type can be found by multiplying the FTP emissions by the ratio for that facility type, speed, and level of congestion.

Overall emissions for a local area will require the user to develop a distribution of VMT on various facility types in the region. Modeling to determine the impact of new highway projects can use expected travel demand for a particular roadway type to get improved estimates of emissions for the existing and the proposed roadway configurations. The use of facility-specific correction factors should improve the calculations of emissions for different roadway types in MOBILE6. This addresses Issue 13 raised in the GAO (1997) report, although further disaggregation might be needed. Other approaches, including the use of emissions rates in units of grams per second or fuel based emissions rates (e.g., in units of grams of emissions per gram of fuel used) could provide a better approach to emissions modeling, particularly at low speeds.

START EMISSIONS

Cold engines and cold catalysts have much higher emissions than those for normal operating temperatures. This makes cold starts a critical com ponent to account for in the model. As noted in Equation 3–1, FTP emissions in grams per mile are based on the weighting of the cold-start, hot-start, and stabilized portions of the FTP.

TABLE 3–4 Parameters for New Facility-Specific Driving Cycles, with LA4 Cycle Used in FTP Included for Comparison

Roadway Classification (LOS Indicates Level of Service)[a]	Average Speed (mph)	Maximum Speed (mph)	Maximum Acceleration (mph/s)	Cycle Time (s)	Cycle Distance (miles)
LA4 Cycle Used in FTP	19.6	56.7	3.3	1,372	7.5
Freewa77y, High Speed	63.2	74.7	2.7	610	10.72
Freeway, LOS A-C	59.7	73.1	3.4	516	8.55
Freeway, LOS D	52.9	70.6	2.3	406	5.96
Freeway, LOS E	30.5	63.0	5.3	456	3.86
Freeway, LOS F	18.6	49.9	6.9	442	2.29
Freeway, LOS "G"	13.1	35.7	3.8	390	1.42
Freeway Ramps	34.6	60.2	5.7	266	2.56
Arterials/ Collectors LOS A-B	24.8	58.9	5.0	737	5.07
Arterials/ Collectors LOS C-D	19.2	49.5	5.7	629	3.36
Arterials/ Collectors LOS E-F	11.6	39.9	5.8	504	1.62
Local Roadways	12.9	38.3	3.7	525	1.87
Non-freeway area wide	19.4	52.3	6.4	1,348	7.25

[a]Level of service (LOS) is a measure of traffic congestion. According to the Transportation Research Board, LOS A has the least congestion and LOS F has the most congestion. LOS G was created to define a "subset of LOS F driving under the worst conditions routinely observed." Source: EPA 1999j.

MOBILE5 computes separate results for each phase and users can supply weightings for each phase, which might be different from the FTP weightings shown above. User-supplied weightings can be used to account for differences in start activity in a local region. However, such accounting is not satisfactory in many applications, such as air-quality modeling, because it allocates the start emissions in terms of grams per mile to an entire trip as opposed to the specific start location. Start emissions was the third issue identified in the GAO report (GAO 1997) on the MOBILE model.

Recall that MOBILE subdivides vehicle classes into technology groups. Each technology group is treated separately in the analysis of start emis sions in MOBILE6. The following discussion refers to obtaining sets of regression equations. Such equations are obtained for each technology group. In this way, MOBILE models the differences in the vehicles in the fleet. In addition, MOBILE6 divides the fleet into normal and high emitters for modeling the effects of inspection and maintenance. (This is a change from MOBILE5 where the fleet was divided into four emissions regimes: normal, high, very high, and super.) Separate regression equations must be developed for each regime. Because of the large number of combinations of regimes and technology groups, a large number of vehicle tests is required to ensure that each group is accurately represented in the model.

EPA plans to use a new start methodology for MOBILE6 (EPA 1999h). In the new scheme, start emissions will be based on the difference between the emissions measured in the cold-start (or hot-start) phase and the emissions measured over the same 505 second driving cycle without a start. (The cold-start and hot-start emissions measured by this process are subsequently adjusted for soak time in MOBILE6, as described below.) To characterize these emissions, EPA obtained data for 77 vehicles tested over the 505 second cycle with no start (EPA 1999k). EPA called this test the hot-running 505 cycle. These vehicles were also tested using the full FTP. EPA developed regression equations for this small vehicle data set, which related the hot-running 505 cycle emissions to individual components of the FTP. These regression equations could then be used to compute the hot-running 505 cycle emissions for the large database EPA has on vehicles with conventional FTP tests. For 1981 to 1993 model-year vehicles, for example, there were 4,416 passenger cars and 1,205 trucks in the database for FTP emissions results.

The cold-start emissions are then found, for each vehicle in the database, as the difference between the cold-start phase (Bag 1) emissions and the hot-running 505 cycle emissions. (The latter is found by regression based on the 77-vehicle data set.) The units of the start emissions are grams per start. The resulting data on emissions as a function of mileage are then used to develop regression equations. EPA did not provide any statistical results for the resulting regression equations.

The regression equations give the start emissions for a vehicle, which has reached the ambient temperature. The FTP requires an engine off time (soak time) of 12 hr to reach this tempeture. MOBILE6 will use a modification of relations developed by the CARB to account for the effect of soak time on start emissions (CARB 1996a). CARB made measurements of start emissions for various soak times and developed a regression equation for the start emissions ratio, defined below, as a function of soak time.

$$\text{Start emissions ratio} = \frac{\text{Emissions at given soak time}}{\text{Emissions at 12 - hr soak time}} \tag{3–5}$$

EPA developed a modified version of the CARB relationship for the start-emissions ratio to be consistent with EPA data on hot starts and cold starts.

The approach to start emissions proposed for MOBILE6 allows for an improved modeling of these emissions. In MOBILE5, such emissions were counted as part of the gram-per-mile emissions associated with overall vehicle operation. Having a separate model for start emissions provides a more accurate representation of these emissions. In addition, the separation of start emissions allows such emissions to be identified with a particular location where the emissions occur. This should provide more accurate results for air-quality models.

Further improvements are possible. A recent review of EMFAC7G (Pollack et al. 1999a) has shown problems with the CARB start-emissions ratio used by EPA. These problems relate to activity data on the number of trips per day (and hence number of starts) and the small data set used to characterize start emissions as a function of the length of time the vehicle has been turned off. Future data collection efforts should include measurements with the hot-running 505 cycle. This will increase the database for measured start emissions. Additional studies should be done to improve the relationship between soak time and start emissions. Such studies should be designed to develop an understanding of the effect of soak time on start emissions at non-FTP temperatures. It is known that VOC and CO emissions increase dramatically at temperatures below about 55°F and that the start mode accounts for much of this increase.

IN-USE DETERIORATION

Even with reasonable maintenance, vehicle emissions increase with mileage and age due to deterioration of engine and emissions-control components. In MOBILE5 and previous versions, EPA has projected major increases in in-use deterioration after 50,000 miles. For instance, MOBILE5 had a nearly 10-fold increase in VOC emissions over certification levels at 100,000 miles. The data analysis for this so called "dog leg" or "kink" effect has been questioned (Sierra Research 1994b). The GAO (1997) report identified deterioration above 50,000 miles as the eighth issue of concern for MOBILE. The substantial deterioration of emissions-control performance has lead to a tightening of certification standards, an increase in warranty requirements, and a requirement for local I/M programs. EPA has made a major attempt to better characterize in-use deterioration and its effects on emissions in MOBILE6. Yet, questions remain about the accuracy of these newly revised deterioration rates.

Issues with In-Use Deterioration

Automobile manufacturers have made major strides in the last decade or so to reduce in-use deterioration of vehicle emissions-control equipment. Improvements in catalyst longevity, replacement of carbureted systems with fuel injection, conversion to platinum spark plugs, and the introduction of closed loop air-fuel ratio control are among the advancements that clearly have improved the longevity of emissions-control performance. These events have motivated EPA to consider major revisions to in-use deterioration factors in MOBILE6.

EPA primarily has relied on an FTP data set it collected from public vehicle-recruitment programs as well as an even larger FTP database supplied by automobile manufacturers to make adjustments to in-use deterioration in MOBILE6. Concerns about the representativeness of these data have been raised. As mentioned previously, EPA's public vehicle-recruitment programs had acceptance rates typically less than 25%, raising concerns about recruitment bias. The automobile manufacturer's database represented relatively young, high-mileage vehicles. This raised concern that more typical aging might result in higher deterioration rates. Other issues include effects of deterioration on start emissions and the overall concern that the FTP based cycle is not representative of more contemporary driving cycles. Effects of deterioration may not be evident under the relatively low-load of the FTP.

Adjustments to In-Use Deterioration in MOBILE6

EPA has addressed issues regarding in-use deterioration in MOBILE6 by using all available new FTP data, splitting these data into start and running modes, and adjusting these data to IM240 tests (EPA 1999l; EPA 1999m). IM240 programs provide large samples of emissions test data using a test cycle that better emulates actual on-road driving compared to other I/M tests. Concerns about recruitment biases in the FTP data, discussed previously, motivated EPA to use IM240 data to correct deterioration estimates computed from FTP data. This adjustment required numerous assumptions and data conversions and introduced uncertainties in the estimates. However, it was felt that the end result was an improvement in estimating deterioration of the real-world fleet.

MOBILE6 In-Use Deterioration Compared to MOBILE5

Figures 3–6, 3–7, and 3–8 present comparisons of MOBILE5 and proposed MOBILE6 VOC running-emissions factors for Tier 0 vehicles (EPA 1999l). As can been seen, deterioration rates for MOBILE6 are substan tially reduced after 50,000 miles with the "kink" eliminated. Newer model years show greater reduction in in-use deterioration than older models.

Conclusions on In-Use Deterioration

EPA has made major changes to lower in-use deterioration in MOBILE6 in response to new FTP data and IM240 data that indicate newer model vehicle emissions-control systems are substantially more durable than older models. However, concerns remain about the new proposed MOBILE6 in-use deterioration rates. These include a small database for start emissions, representativeness of the FTP database (and even I/M databases, particularly for high emitters and typical aged vehicles), and overall accuracy of the FTP-based cycle to represent current real-world driving cycles. Because of these concerns, EPA should establish a long-term testing program to characterize the in-use deterioration of representatively aged new-technology vehicles using a driving cycle more representative of real-world driving conditions. It should also be noted that lower in-use exhaust deterioration rates might increase the discrepancy between MOBILE6 emissions estimates and emissions estimates obtained through tunnel and ambient studies. Chapter 4 describes these field observations in detail.

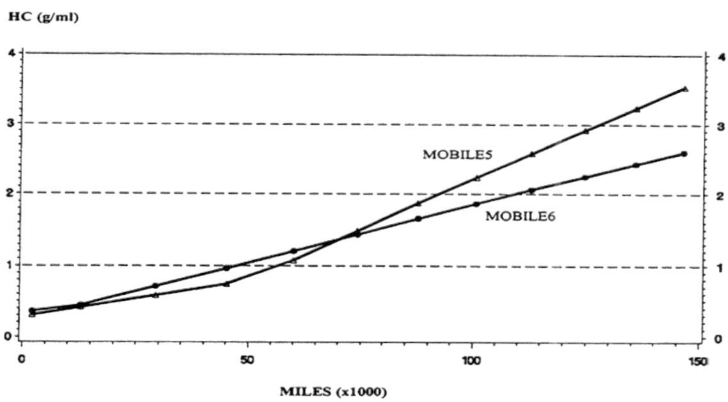

FIGURE 3–6 Comparison of MOBILE5 and MOBILE6 hydrocarbon (HC) emissions factors as a function of mileage for 1981 model-year passenger cars. Note that this report usually uses the term VOCs as opposed to HCs to refer to the general class of gaseous organic compounds. Source: EPA 1999l.

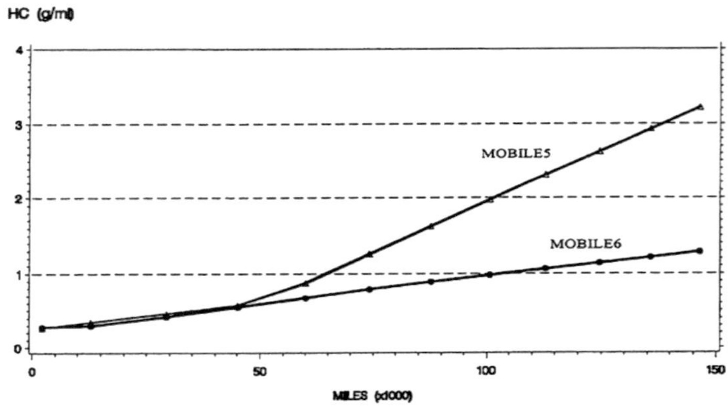

FIGURE 3–7 Comparison of MOBILE5 and MOBILE6 hydrocardon (HC) emissions factors as a function of mileage for 1987 model-year passenger cars. Source: EPA 1999l.

INSPECTION AND MAINTENANCE ISSUES

Emissions Estimates and Assumptions for I/M Programs

The use of the MOBILE model for the calculation of the benefits of vehicle I/M programs is one of the most controversial applications of the model. This was identified as Issue 9 in the GAO (1997) report on the MOBILE model. Individual nonattainment areas receive emissions-reduction credits in their State Implementation Plans (SIPs) based on the predictions of the MOBILE model. The current version of the model, MOBILE5, generally gives the most credit for a centralized inspection procedure (i.e., one in which the testing station does not do repairs) using a dynamometer test known as IM240. The IM240 test uses a driving cycle, which is a reduced portion of the FTP cycle, to better characterize real-world emissions compared with a 2-speed idle test of the vehicle.

Various studies of I/M procedures have questioned the effectiveness of such programs, especially the benefits predicted by MOBILE5. The National Highway System Designation Act, enacted in 1995 allowed states to provide alternative methods for determining the benefits of these programs for their SIPs, provided they could adequately demonstrate the benefits of such programs.

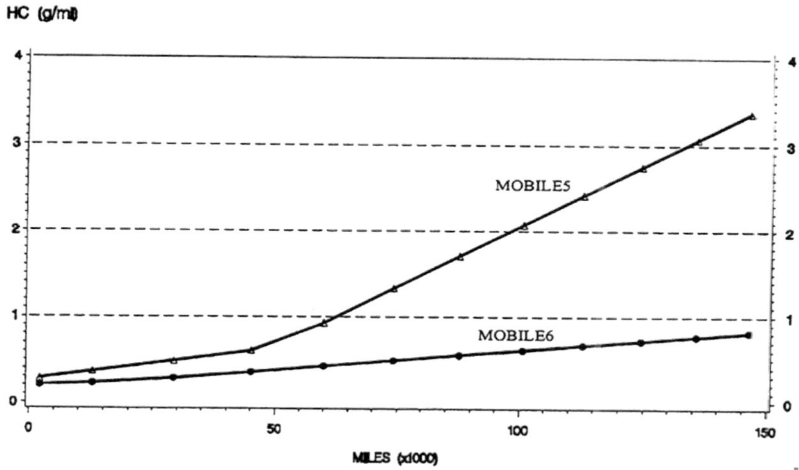

FIGURE 3–8 Comparison of MOBILE5 and MOBILE6 hydrocarbon (HC) emissions factors as a function of mileage for 1992 model-year passenger cars. Source: EPA 1999l.

The effectiveness of I/M programs relies on a combination of technical and behavioral effects. The actual inspection program uses a short test that measures the vehicle's emissions. The short test is necessary because an FTP-type test would be too costly and too time consuming. The short test is designed to identify vehicles that have high emissions. The cut points for the tests—the emissions results that identify vehicles as failing the test—are set so that there is very little chance that a vehicle meeting the standards will fail the test. This means that there is a significant probability that some vehicles that do not meet the standards will pass the test. Further, vehicles that fail are repaired only to the extent that they will pass the short test. Thus the cut points might affect the repair effectiveness.

Other exhaust tests are possible in addition to the IM240 test. The simplest exhaust emissions test measures idle emissions. This is usually supplemented by a measurement with a no-load engine speed of 2,500 RPM. This test (or test combination) does not require a dynamometer. An alternative test currently used is known as the acceleration-simulation mode or ASM. This test uses a single engine load to simulate a particular vehicle operating point. Because the dynamometer that is used for this test does not have to simulate the instantaneous changes required for the IM240 test, a less costly dynamometer can be used.

In addition to the exhaust emissions test, tests are also done on the evaporative control system. It is also possible to conduct tests for tampering[1] as part of an I/M program, which are also done in addition to the exhaust emission test.

The technical assessment of the effectiveness of I/M programs requires an assessment of the rate at which vehicles fail the short test (alternately called the failure rate or identification rate) and the effectiveness of their repair. The model must also account for the effect of a cost cap that governs repairs made to vehicles failing inspection. If the vehicle's repairs cost more than a predetermined amount, additional repairs need not be performed on the vehicle. The CAAA90 set this cost cap in 1990 at $450 for late-model vehicles, which increases with inflation.

The behavioral aspects of I/M programs can also have an impact on their effectiveness. Individuals who know that their vehicles will be inspected might maintain their vehicles better. This would result in a lower failure rate than predicted and in lower emissions that might not be credited to the program. It is also possible, particularly in decentralized programs where test and repair are combined, for some inspectors to fraudulently pass a vehicle that should otherwise fail. One common problem with older vehicles was that individuals could adjust a repaired vehicle so that its performance would improve, but its emissions would increase. This is less of a problem with modern, computer-controlled vehicles. Finally, owners of vehicles that are subject to inspection might avoid inspection completely or their vehicles might never be repaired after an initial failure. MOBILE attempts to capture some of these effects by using input data for such vehicles that have not been inspected.

The analysis of I/M programs is based on the division of the fleet into emissions regimes. MOBILE6 uses two regimes, normal emitters and high emitters, whereas, previous versions of MOBILE used four regimes, normal, high, very high, and super. Vehicles are assumed to move from a lower-emitting regime to a higher-emitting regime as a result of some failure of an emissions component. Detection and repair of the failed component moves the vehicle from a higher-emitting regime to a lower-emitting one. This component failure is contrasted to the normal deterioration that is expected for well-maintained vehicles.

The benefits of I/M have been computed in a separate model known as the TECH model. The results of this model are used as inputs to MOBILE. Starting with MOBILE6, the TECH model will no longer be

[1]Tampering as defined in the MOBILE model is the malfunctioning of one or more emissions- control device due to either deliberate disablement or mechanical failure.

used. The calculations previously done in this model are integrated into MOBILE6.

The I/M benefits for running emissions are computed as follows in MOBILE6 for a program using IM240 (EPA 1999n):

1. All emissions benefits come from identifying and repairing high-emitting vehicles.
2. The identification (failure) rate for high emitters is determined by regression equations, which are functions of the cut points used in the IM240 tests.
3. High-emitter emissions are independent of vehicle age or mileage. Repaired emissions are computed as a multiplicative adjustment factor times the normal emissions level. The normal emissions, and hence the repaired emissions, are a function of vehicle mileage.
4. The multiplicative adjustment factor used in Step 3 is a function of vehicle age. This factor is found from regression equations based on data from the Arizona IM240 inspection program.
5. The after-repair emissions are adjusted by further multiplicative correction factors to account for more stringent cut points and for technician-training effects.
6. The final emissions after the I/M process are the sum of the following components:

   - the emissions of the normal fraction of the vehicle fleet, that is assumed not to be inspected or repaired;
   - the emissions of the high-emitters fraction of the vehicle fleet that is not identified and repaired;
   - the emissions of high emitters that have been identified and repaired;
   - the emissions of waivered vehicles, such as high emitters that have been identified but not fully repaired because of a cost limit, or older vehicles not required to be tested; and
   - disappearing" vehicles that never show up for their emissions test or fail an initial test and never receive full repairs or a cost waiver.

A similar analysis is applied to start emissions as well. Because I/M exhaust measurements assume a fully warmed vehicle, start operations are not tested. The identification and repairs to reduce start emissions are assumed to occur through the identification and repair of failures in running emissions.

The analysis outlined above is done on a year-by-year basis. MOBILE has fixed mileage accumulation rates (for a given vehicle class) so that there is a unique relationship between the average age and average mile age accumulation in MOBILE. The year-by-year emissions benefits for a given starting year are then combined to account for deterioration between inspection cycles. The resulting process is called the sawtooth method because of the appearance of the resulting graph. Emissions drop after inspection and repair. They then are assumed to increase due to normal vehicle deterioration until the next inspection cycle. This process is repeated over the lifetime of the vehicle.

Modifications to this process are made when the I/M program calls for off-cycle inspections. Such inspections might be required upon a change of ownership or by the use of remote-sensing detectors to identify high-emitting vehicles that can be called in for inspection.

Several researchers have questioned the magnitude of the emissions benefits estimated for I/M programs in the MOBILE model. These studies are described below. Much of this criticism has been based on the data collected using remote-sensing devices (RSD). These devices measure the concentration of emissions in the exhaust as the vehicle is passing by a detector. This method has the potential advantage of making measurements with actual on-road driving. However, the vehicle is measured under a single operating mode and that mode might differ as the RSD is moved from location to location.

Early remote-sensing studies (Stedman 1989; Lawson et al. 1990) and one study based on ambient measurements (Scherrer and Kittelson 1994) found very little change in on-road fleet emissions that could be attributable to I/M programs. Wenzel (1999) has analyzed extensive data from remote-sensing measurements of vehicles in the Arizona I/M program. Over 450,000 vehicles with both IM240 and remote-sensing measurements were available for his analysis. He found that the emissions from vehicles measured within 1 month after their inspection and maintenance had CO emissions[2] reductions of 12%. This was less than the estimated IM240 emissions reduction of 14.5% and the MOBILE predictions of 16%. Further, Wenzel found that the improvement decreased over time. Vehicles measured within 3 to 6 months after I/M had only a 9% reduction; within 12 to 15 months the reduction was only 6%. This study shows a deterioration of vehicles after repair is much greater than the deterioration of normal-emitting vehicles. This finding contradicts a fundamental assump

[2]Wenzel's study focused on CO emissions because no NO_x results were available from the remote-sensing readings and there was poor correlation between remote sensing and IM240 VOC data for 1991 and later model-year vehicles. All the data reported here are on the emissions improvement in the overall fleet. Improvement data are often reported as the percentage of emissions reduction in failed vehicles only. This percentage is much higher than the overall fleet reduction.

tion in MOBILE that both normal emitters and repaired vehicles have the same deterioration rate. Wenzel has not used the data on vehicle speed to select out vehicles, which might show high emissions due to normal enrichment operation even though the vehicle is operating properly. Jimènez (1999) has shown that there is a distribution of engine power conditions at remote sampling locations, which might cause this erroneous indication of a high-emitting vehicle. It is not clear how this would affect the emissions results percentages obtained by Wenzel.

Wenzel also analyzed the effects of vehicle load on the IM240 emissions reduction. He noted that RSDs were placed in locations where the vehicle would be under moderate load to obtain a strong emissions signal. He determined the emissions reductions for the moderately loaded portions of the IM240 cycle and found that the measured CO reductions for the moderately loaded portions of the cycle were only 77% of the measured reductions for the entire cycle. Thus, one possible discrepancy between the emissions reduction for the remote-sensing and the IM240 results might be due to differences in operating conditions.

"Disappearing" Vehicles

I/M programs maintain detailed records on vehicle emissions tests. From these records it is possible to determine the number of times a failed vehicle is retested and the change in the reinspection emissions resulting from repair of a failed vehicle. Recent studies have shown that a significant fraction of failed vehicles never appear for a retest (Wenzel 1999). For example, an EPA study of the I/M program in Arizona (EPA 1997c) found that 15% of the failures has not been retested. In the draft plans for the MOBILE6 analysis of I/M programs, EPA staff stated that the default value for the noncompliance rate will be 15%. However, EPA has a standard that requires enhanced I/M programs in certain nonattainment areas to achieve a noncompliance rate of 4% or less. The noncompliance rate includes both vehicles that disappear after an initial I/M failure and vehicles that never show up for an I/M inspection. According to EPA (1999n), this is a "generous default" because EPA staff analysis showed actual rates greater than 20%. Users can set higher rates, based on actual data, when they run MOBILE.

The discussion of the noncompliance rate (EPA 1999n) demonstrates the conflict that sometimes appears in the MOBILE model. Although the analysis of I/M programs shows a noncompliance rate greater than 20%, EPA staff has selected a default rate of only 15%. In addition, the discussion notes that this choice "does not constitute a policy by EPA to allow the use of this value for SIP purposes." If MOBILE6 does not reflect the best information available, its value and accuracy will continue to be questioned by the user community.

Estimating I/M Repair Effects

The repair effectiveness part of MOBILE5 has been criticized (Harrington et al. 1998) because it is based on the relatively small number of 266 vehicles. All of these vehicles were repaired in a laboratory setting instead of actual repair shops. In addition, it was not possible to repair all the vehicles to pass the IM240 test; 46% of the vehicles had emissions above the cut points even after repairs. In these cases, EPA extrapolated the repair results observed to determine what the emissions results would be, if additional repairs could be performed.

For MOBILE6, EPA has based repair effectiveness on the results of the Arizona I/M program for 1981–1993 vehicles. The ratios were computed by combining data for cars and trucks and sorting the data by age. For each age group, the ratio of mean emissions from failed and repaired vehicles to the mean emissions from passed vehicles was calculated. These ratios were then regressed against age to get the regression equations for the ratio of after-repair emissions to the emissions of normal emitters used in MOBILE6.

Because the Arizona data were only for one particular set of cut points, EPA then used data from the repair database that was used for MOBILE5 to determine the effect of cut points on after-repair emissions. These show significant effects for VOCs and NO_x. Compared with the reference cut points of 1.2 g/mi for VOCs, reducing the VOC cut point to 0.4 g/mi is predicted to reduce after-repair emissions by a factor of 0.59. Similarly, reducing the NO_x cut point from its reference value of 3.0 g/mi to 1.0 g/mi is predicted to reduce after-repair emissions by a factor of 0.489. For the range of CO cut points reported (20 to 15 g/mi) the multiplicative factor is 0.87.

Estimating Technician-Training Effects

MOBILE5 has credits that increase the benefits from an I/M program for areas that have programs to train repair technicians. The repair-effectiveness data for MOBILE6 are assumed to be those achieved by master technicians. MOBILE6 will thus increase the estimated after-repair emis sions for areas that have no training program. These emissions increases will be 78% for VOCs, 174% for CO, and 39% for NO_x. These significant increases are based on a single study in which 11 mechanics participated, repairing three cars each. The emissions comparison is based on the difference between the repair results of the students and those of the trainer. No data were available to directly compare the differences between the students' performance before and after the training, which is the factor to be quantified. EPA acknowledges that the training effect in MOBILE6 is based on "limited data." Users in areas that do not have a technician-training program can enter the expected increase in emissions due to the absence of a training program. Such an estimate might be based, for example, on engineering judgment or data from other programs.

It is reasonable to assume that I/M programs that train mechanics should be able to obtain a greater emissions reduction from their programs compared with programs without technician training. However, the accounting for this training effect in MOBILE is based on minimal data of questionable applicability. Additional data are required to make a direct connection between a particular training program and the related improvement in emissions reduction from repair under an I/M program.

Tampering Effects

MOBILE5 has data on effects that expresses tampering rates as a function of vehicle mileage. Separate tampering rates are available for different components. These tampering data predict that the incidence of tampering will be much less in later- model vehicles. Harrington et al. (1998) examined the incidence of tampering in the Arizona I/M data. They based their evaluation on the amount of tampering in failed vehicles rather than the amount of tampering in the fleet. Consequently, they were unable to make a direct comparison with the tampering data used in MOBILE. For 1995 and 1996 model-year vehicles, the latest year in their study, they found that 6 of the 16 failing vehicles had been tampered with. Based on this very small sample, of failing vehicles only, the authors concluded that the tampering might account for a large fraction of failures in late-model vehicles. Additional data are required to determine the true extent of such tampering. The committee is aware that the California Bureau of Automotive Repair is preparing a report on a large data set from vehicles that were randomly pulled over and checked during 1997–1999. This report, which was not publicly released and therefore not available to the committee, should provide useful data on the true extent of tampering in the on-road fleet.

On-Board Diagnostic Effects

MOBILE6 will consider the effects of electronic diagnostic systems on-board the vehicle. On-board diagnostic (OBD) systems were required on all cars starting in the 1996 model year. Such systems have a malfunction indicator light (MIL) that is supposed to be illuminated when the diagnostic system detects a malfunction that would increase the exhaust emissions to 1.5 times the applicable standard or more. EPA assumes that a check of the OBD system will be a part of the I/M tests.

In modeling OBD systems, EPA assumes that such systems will detect 85% of the high-emitting vehicles (EPA 1999o; EPA 1999p). In other words, the MIL will illuminate in 85% of the vehicles that are high emitters. The response rate of drivers to the illuminated MIL is assumed to depend on the vehicle mileage and the presence of an I/M program. Table 3–5 shows the response rates. It is assumed that drivers will not respond to the illuminated MIL unless there is an I/M program in place or the vehicle repair is still under warranty.

EPA believes that it might be possible to use an I/M procedure in some future year —they suggest 2001—that has no exhaust emissions measurement, only an OBD check. However, they recognize that this is an unproven concept, and MOBILE6 will have calculation procedures that can handle OBD both as a stand-alone system and a system used in conjunction with an exhaust emissions measurement. Figure 3–9 displays the impact of the OBD only and OBD in conjunction with an I/M program on deterioration rates for Tier 1 LDVs and LDTs.

The information presented in Table 3–5 illustrates two critical issues with MOBILE. One is the need to predict the life-cycle performance of new emissions control systems and the other is to develop greater knowledge of how motorists respond to factors such as illumination of MIL. The assumed response rates to MIL lights are reasonable guesses; however, model results developed with these response rates could be used to justify-the effectiveness of I/M programs.

TABLE 3–5 Response Rate to Illuminated Malfunction Indication Light

Mileage Range	Warranty Coverage for Mileage Range	Response Rate in I/M Areas	Response Rate in non-I/M Areas
0–36,000	Full warranty coverage	90%	90%
36,000–80,000	Only catalysts and electronic control module under warranty	90%	10%
Over 80,000	No warranty coverage	90%	0%

Source: EPA 1999o; EPA 1999p.

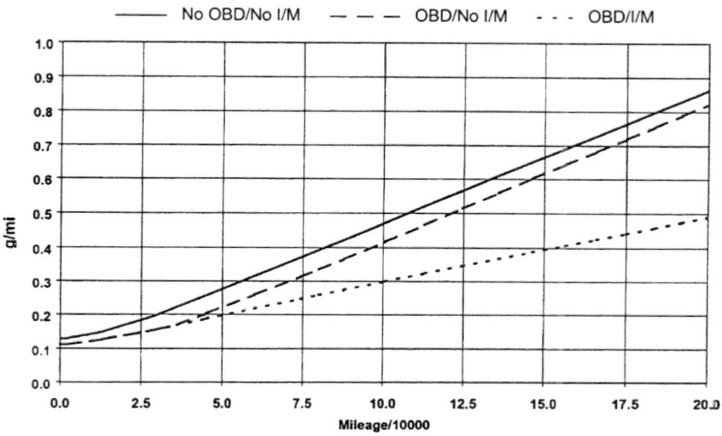

FIGURE 3–9 Deterioration rates for nonmethane hydrocarbons (NMHCs) for Tier 1 light-duty passenger cars and class 1 light-duty trucks as a function of OBD and I/M. Note that this report usually uses the term VOCs as opposed to NMHCs to refer to the general class of gaseous organic compounds. Source: EPA 1999p.

Summary on Inspection and Maintenance

The modeling of I/M programs has been one of the most controversial elements of the MOBILE models. The changes proposed for MOBILE6 are unlikely to resolve any of the significant issues in these controversies. Many vehicle owners are apparently able to avoid inspection or fail to get their vehicles repaired. Such vehicles must be accounted for properly in MOBILE. Remote-sensing measurements and IM240 data indicate higher deterioration rates for repairs to failed vehicles than those used in MOBILE. This might be due to mechanical problems with the vehicles or to tampering by owners. It seems intuitively obvious that vehicle inspection and maintenance should result in cleaner, lower-emitting vehicles. This basic intuition has not been supported by unambiguous data on the emissions reductions and cost-effectiveness of I/M programs. Early indications are that MOBILE6 will substantially reduce the emissions-reduction benefits from I/M compared with MOBILE5 (Clean Air Report 1999).

EPA appears to be heading towards the use of OBD systems as an alternative to current I/M programs. These systems are untried as an I/M tool and their treatment in MOBILE6 appears to be based on little more than assumptions about system effectiveness and driver behavior. It is not clear that any reasonable model will be able to capture all the effects of I/M—including the assumption that vehicle owners in I/M areas will maintain their vehicles more frequently—without a significant testing program. The testing program would be necessary to determine the model parameters and validate the results of the model.

AIR-CONDITIONING EFFECTS

Vehicle air conditioners place an additional load on the engine, as described in Equations 3–2 and 3–3. This load increases both the fuel consumption and the emissions for a given vehicle speed and road load. This emissions increase is an issue for both vehicle certification and emissions inventory. Emissions from vehicles with operating air conditioners was the fourth issue raised in the GAO (1997) report on MOBILE.

Starting with model-year 2000 vehicles, EPA is phasing in a new certification procedure for vehicles with air conditioners. Previously, the load on the dynamometer was increased by 10% to account for the effect of air-conditioning during certification. The air conditioner is not actually operated during this certification test. The new certification procedure has a separate test that measures emissions from vehicles with the air conditioner operating. The results of this test will not be used in MOBILE, however.

EPA has added a new set of calculations to determine emissions from air-conditioning operation in MOBILE6.[3] There are two elements to these calculations: (1) the determination of emissions for full-load air-conditioning operation and (2) the estimation of the amount of actual air-conditioning use. These effects were measured for actual driving cycles, not for the certification cycle to be used in the SFTP.

The effect of full air-conditioning operation was determined from a sample of 37 vehicles-23 passenger cars and 14 light-duty trucks-from model years 1990 to 1996 (EPA 1998e). The data were analyzed for both normal and high emitters, but only five of the vehicles were high emitters for at least one pollutant. All vehicles were tested over 15 different driving cycles. These included the speed-correction factors and facility driving cycles described for determining the effects of actual in-use operation on emissions. Additional cycles used in this study are shown in Table 3–6.

[3]MOBILE5 does contain an option for air-conditioning calculations. However, the data for these calculations were obtained in the early 1970s and users were advised not to use this calculation option.

TABLE 3–6 Parameters for Additional (Nonfacility-Specific) Driving Cycles

Cycle Name	Average Speed (mph)	Maximum Speed (mph)	Maximum Acceleration (mph/s)	Cycle Distance (miles)
New York City Cycle	7.1	27.7	6.0	1.18
CARB "Unified" (LA92) Cycle	24.6	67.2	6.9	9.81
Start Cycle (ST01)[a]	1.39	20.2	5.1	1.39

[a]The ST01 start cycle was an earlier version of the ST03 cycle used for air-conditioning tests in the SFTP.

Source: EPA 1998e.

The results of these tests will be represented in MOBILE6 as ratios of emissions with air-conditioning to emissions without air-conditioning. For running emissions these ratios are presented as regression equations that are quadratic functions of mean cycle speed. The input data to these regressions are the emissions ratio for each cycle and the mean cycle speed. These regression equations have r^2 values between 0.25 and 0.80 depending on the pollutant, vehicle type, and emitter class. For start emissions, a single value is used to give the ratio of emissions with air-conditioning to those without air-conditioning.

The effect of air-conditioning ranges from almost no change to more than a doubling of emissions. The effect is most pronounced at lower speeds and is generally higher for passenger cars than for light-duty trucks. The emissions ratios discussed above apply only when the air-conditioning compressor is at maximum operation.

MOBILE6 also accounts for the amount of air-conditioning operation. The activity data for air-conditioning operation were obtained on a fleet of 20 vehicles operating in Phoenix, Arizona, from August to October 1994 (EPA 1998f). Although the actual air-conditioning load depends on the torque generated by the compressor, no data were available for this variable. Consequently, all data on the extent of air-conditioning load were measured in terms of the fraction of time that the air-conditioning compressor was actually operating. This was called the compressor-on fraction.

The compressor-on fraction was modeled as a quadratic function of the heat index. The latter variable is intended to account for the combined effect of temperature and humidity on human comfort. Three separate regression equations were used to account for differences in solar insulation during different diurnal periods:

- nighttime, defined as sunset to sunrise;
- peak sun, defined as noon to 4 p.m.; and
- morning or afternoon, defined as the other daytime periods.

Because of small samples for high and low values of the heat index, the regression equations produced counterintuitive results in these regions. For example, the nighttime equation produced greater activity than the daytime equation. Because of this, additional regression equations were derived for all daytime data and for the entire data set. The MOBILE6 air-conditioning activity model selects the regression equations to be used based on the heat index. In general, the equations for different periods are used for the middle range of the heat index. When counterintuitive results occur from these regression equations, the combined equations are used. The compressor-on fraction is assumed to be zero for a heat index of 65°F and below and 1 for a heat index of 110°F and above. The MOBILE6 model will allow user input on cloud cover. Cloud cover is accounted for by using the nighttime equations during periods of cloud cover.

The MOBILE6 model will also have data on the fraction of vehicles equipped with air conditioners and the fraction of air conditioners that are malfunctioning. These data will be expressed as a function of model year. The air-conditioning corrections will only be applied to the fraction of vehicles that have air conditioners that are installed and operating.

During the development of the model for compressor-on fraction, consideration was given to the effects of vehicle speed, soak time, trip duration, and fraction of idle time during the drive. None of these effects was found to be significant within the limitations of the data available.

The consideration of air-conditioning effects provides a good illustration of the uncertainty and resource issues in mobile-source modeling. Although the results of the proposed air-conditioning submodel are subject to inaccuracies and inconsistencies because of data limitations, they appear to provide a reasonable estimate of an important effect on emissions. Improvements in the accuracy of the air-conditioning effect will require additional data. Without some estimate of the uncertainty due to air-conditioning effects compared with other uncertainties in the model, it is difficult to decide whether more resources should be spent on improving the air-conditioning results or on improving other parts of the model. Finally, the uncertainties in this submodel are of two kinds. First, there are the statistical uncertainties, which can be estimated quantitatively by the error terms in the regression equations. Second, there are the unknown uncertainties which are due to factors such as the effect of vehicle speed on compressor-on fraction, the validity of the heat index as a measure of air-conditioner use, individual driver behavior in air-conditioner use, or the effect of the actual compressor torque.

EVAPORATIVE EMISSIONS

Evaporative emissions have several categories. Four main physical mechanisms are used to account for evaporative emissions: diurnal emissions, resting losses, hot-soak emissions, and running losses. The combination of diurnal emissions and resting losses are measured together as "real time diurnal" [RTD] emissions. The measurement and characterization of evaporative VOC emissions is based almost exclusively on certification test procedures with no real-world measurements. New vehicle evaporative emissions certification to the 2 gram per test standard has been based on the sealed-housing for evaporative-determination (SHED) testing of a hot-soak and a 1-hr diurnal emissions test. The new evaporative test procedure introduced in 1996 includes real-time, multiday diurnal, and running-loss emissions measurement. Results of these tests provide a better picture of evaporative emissions, but the test procedure is lengthy and complex, and places a practical limit on the number of vehicles tested.

The database of measurements for the in-use fleet is improving with MOBILE6 using data from approximately 300 vehicles. Hot-soak and running-loss evaporative emissions are highly skewed with substantial contributions from high emitters that apparently relate to liquid leaks. Recent studies (Gorse 1999) indicate that evaporative VOC emissions from the fleet studied exceed tailpipe emissions by a factor of more than 2. This is at variance with the results shown in Figure 3.2. This figure shows an evaporative to tailpipe ratio of about 0.7 for Chicago and New York, using the MOBILE5b approach to calculating evaporative emissions.

Characterization of Multiday Diurnals

Diurnal emissions come from natural cycling of the ambient temperature and the resulting pressure-driven emissions of fuel vapors. The magnitude of these emissions depends on the ambient temperature variation, fuel vapor pressure, and the period of vehicle nonoperation. Resting-losses are gauged by the permeation of fuel through tanks, lines, and fittings, and liquid leaks that are not the result of temperature variation. These occur while the vehicle is not operated and are also captured in the multiday diurnal tests. Figure 3–10 shows the temporal variation of these two types of evaporative emissions for a typical vehicle without large leaks. Real-time diurnal emissions tests of 270 vehicles provide the data for estimating diurnal and resting-loss emissions.

The emissions from the fleet tested are highly skewed, with liquid leakers dominating the high end of diurnal and resting-loss emissions. In MOBILE6, the fleet is divided into normal and high emitters. The diffi culty of correctly determining the distribution of emissions in the vehicle fleet and the level of emissions from the high emitters remains a difficult problem.

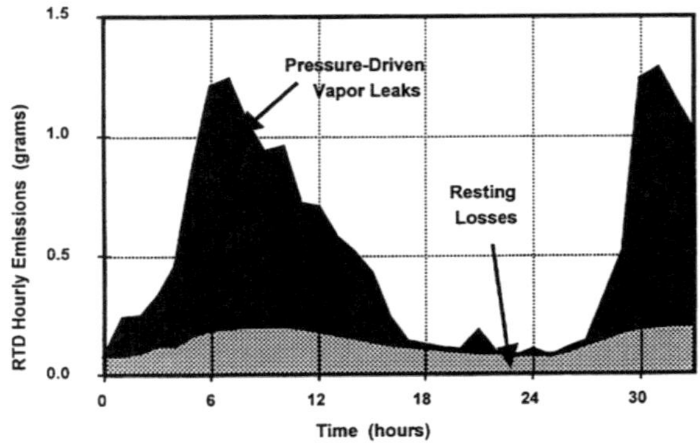

FIGURE 3–10 Illustration of real-time diurnal measurement for a typical vehicle without large leaks showing the combination of pressure-driven losses due to diurnal temperature variation (dark area) and resting losses (light area). Source: EPA 1999q.

Characterization of Hot-Soaks

Hot-soak emissions are evaporative emissions occurring, by definition, during the first hour following engine shutdown. Most of the hot-soak emissions occur during the first 10 min (EPA 1998g). Hot-soak emissions result from heating the fuel above ambient temperatures after vehicle operation, that is, during the hot-soak. They come primarily from the fuel tank and, in carburetor vehicles, from the carburetor bowl. They depend upon vehicle technology, ambient temperature, fraction of fuel in the tank, and fuel vapor pressure. With the replacement of carburetors by fuel injectors, the contribution of hot-soak emissions to evaporative emissions has fallen. Hot-soak emissions are not as skewed as diurnal and resting-loss emissions.

Characterization of Running-Losses

Evaporative emissions occurring during vehicle operation are termed running-losses. These emissions are measured while the vehicle is being operated on a chassis dynamometer. Emissions depend upon driving cycle, fuel vapor pressure, and ambient temperature. The importance of these emissions was not well recognized at the time of the development of MOBILE5. Measurement of running-loss emissions by the CRC from pre-1992 vehicles (McClement et al. 1998) and from 1992 and later vehicles (McClement 1999) show emissions in reasonable agreement with those predicted by MOBILE5 for the newer vehicles but much greater than predicted for the older vehicles. Running-loss emissions data developed for MOBILE6 indicate that MOBILE5 underestimates emissions by a factor of 2 to 4 (EPA 1999r). Because of the skewness of the emissions and importance of the gross liquid leakers, an accurate determination of the true in-use fleet running-loss emissions is difficult.

FUEL EFFECTS

Reformulated Gasoline Effects

Use of RFGs can be specified in MOBILE5 input. The effects of RFG on vehicle emissions are determined mainly by the RVP and oxygenated-fuel parameters, discussed below. The effects of RFG depend upon the season. If RFG effects are modeled for summer conditions, any user-supplied RVP and oxygenated-fuel inputs are overridden by the provisions of the RFG rules. These rules require that for Phase I RFG (1995 through 1999), the maximum RVP be 7.1 and 8.0 pounds per square inch (psi) for fuel-volatility Regions 1 (southern) and 2 (northern), respectively. Phase 2 RFG, which begins in year 2000, requires that all regions have an RVP no higher than 6.7 psi.

In the winter, fuel RVP is not regulated. What is much more important in winter is the fuel oxygen content, and default values might be overridden by user-selected values as long as they are above 2.1 weight percent (wt%). The effect of the fuel sulfur level is fixed in MOBILE5, because RFG is assumed to have a specific sulfur content.

In MOBILE6, there will be some changes in the way that the benefits of RFG are applied. Although MOBILE5 calculates the effect of RVP and oxygen before adjusting for RFG, MOBILE6 (as proposed) would adjust for RVP, oxygen, and sulfur before adjusting for RFG.

Oxygenated-Fuel Effects

Oxygenates are oxygen-containing organic compounds, alcohols, or ethers, that are added to gasoline to achieve enleanment in the combus tion process. Enleanment typically results in a reduction of CO and VOC emissions and an increase in NO_x emissions from the engine. The use of oxygenates (oxyfuels) was mandated by EPA under the CAAA90 in certain regions during winter months in response to violations of the CO air-quality standard. Oxygenates are also a component of the RFG program, which has the objective of reducing ozone-precursor VOC emissions. The potential increase in NO_x emissions and a reduction in fuel economy are the main disadvantages of adding oxygenates to gasoline.

In 1997 the Office of Science and Technology Policy (OSTP) published "Interagency Assessment of Oxygenated Fuels," which reviewed the state of the winter oxyfuel program (NSTC 1997). In the assessment of the air-quality effects of the program, several issues were identified that are relevant to the crediting of such a program in MOBILE5a:

- The observable reduction in ambient CO levels that could be attributed to the use of fuel oxygenates was lower by a factor of 2 or 3 than the amount predicted by the MOBILE5a model.
- The MOBILE5a model predicted CO emissions reductions that were about a factor of 3 larger than other EPA models, notably, a version of the Complex Model developed to represent fuel effects for the on road fleet.
- The emissions database was inadequate to accurately predict the effects of fuel oxygenates on CO emissions at temperatures below about 50°F. The available data indicated that the emissions reduction was decreased at low temperatures compared with the effect at 75°F.
- Because of improvements in emissions-control technology, new vehicles experienced relatively little CO emissions reduction from fuel oxygenates.

According to EPA documentation for MOBILE6 (EPA 1998h), the model will have reduced oxygenate benefits "matching MOBILE6 predictions with ambient CO data." Unfortunately, adequate data do not exist from either ambient air analyses or from vehicle-emissions studies to develop an accurate prediction. For example, there are no accurate measurements of the ambient air effects of oxyfuels for any region in the eastern or mid-western states upon which to base a prediction. The vehicle-emissions data for oxygenate effects, upon which the MOBILE6 analysis is based, are about 10 years out of date, do not represent the current fleet, and are inadequate for temperatures below about 50°F.

EPA documentation does not give a representative prediction for MOBILE6 to compare with the MOBILE5b or earlier model versions, but it does appear to reduce the oxyfuel effects on CO emissions. The OSTP assessment found that one of the reasons for MOBILE's overprediction of the oxyfuel effect was that the model incorporated large reductions for high CO emitting vehicles and used a vehicle distribution with a population of high emitters that was too large. It is not clear that the latter problem has been corrected in MOBILE6.

The results from a recent emissions study conducted by the Colorado Department of Public Health and Environment should be noted. The study was conducted at 35°F on a fleet of 31 vehicles selected as representative of the Colorado on-road fleet and used a 3.5 wt% ethanol fuel. The average CO emissions reduction was about 11%, which is about one-third of the MOBILE5a predicted benefit (Ragazzi and Nelson 1999).

Sulfur Effects

Since the early 1970s, it has been known that sulfur in gasoline affects the conversion efficiency of automobile three-way catalysts. Studies of Tier 0 vehicles indicated that adverse effects of sulfur on catalysts could be substantially reversed if they were subsequently refueled with low-sulfur gasoline (AQIRP 1992). New vehicles are certified on indolene fuel, which typically contains very low sulfur levels in the range of 30–50ppm. In contrast the real-world sulfur levels in the United States average about 330 ppm with peaks in the range of 1,000 ppm. Figure 3–11 presents a distribution of sulfur levels in gasoline for select cities in the United States. Generally, premium grade gasoline is on the low end of the sulfur range because of the refining process used to produce this type of fuel.

The emissions impact of sulfur in gasoline is reflected in EPA's Complex Model. This model, which applies strictly to 1990 technology vehicles, was used by EPA to adjust emissions factors in MOBILE5 to the national average sulfur content of 330 PPM. Appropriate adjustment was also made for Phase I RFG fuel, which assumed an average sulfur content of 220 ppm. Use of MOBILE5 in areas that significantly deviated from the national average sulfur content would have some inaccuracy introduced by not accounting for the sulfur effect. The basic effects of sulfur in gasoline on Tier 0 vehicles reported by EPA are summarized in Table 3–7.

A considerable amount of new data on the impact of sulfur in gasoline has been generated in the last few years, driven to some extent by substantial concerns about larger impacts on new LEV and beyond vehicle technologies. EPA has analyzed this information and has developed a much more accurate estimate of fuel sulfur effects. EPA (1999s) proposes to incorporate this information in a sulfur-emissions correction in MOBILE6, which will allow users to input the area-specific sulfur content of gasoline. This will greatly improve the simulation of gasoline sulfur factor impacts. Impacts will be segregated into start and running emissions. Figures 3–12 through 3–15 provide an indication of the composite correction factors for a range of vehicle technologies.

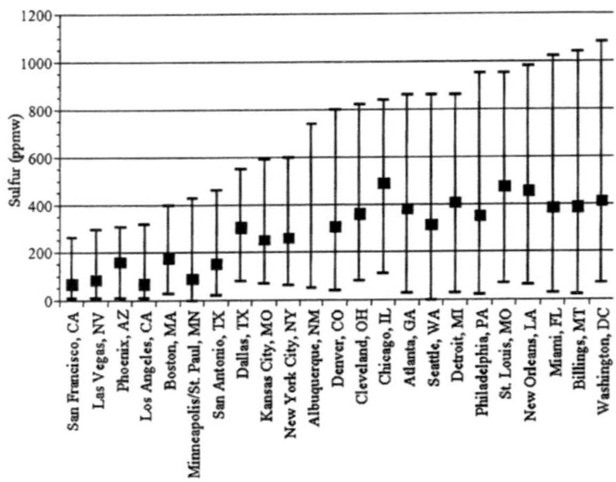

FIGURE 3–11 1996 annual average and extreme sulfur levels in gasoline for selected U.S. cities. Source: Darlington et al. 1999.

TABLE 3–7 Estimated Emissions Reductions Due to Reductions in Fuel Sulfur for a Normal-Emitting Tier 0 Vehicle

Reduction in Emissions, % (400 ppm to 50 ppm Sulfur)		
NO_x	VOC	CO
6.3%	18.8%	21.7%

Source: EPA 1999s.

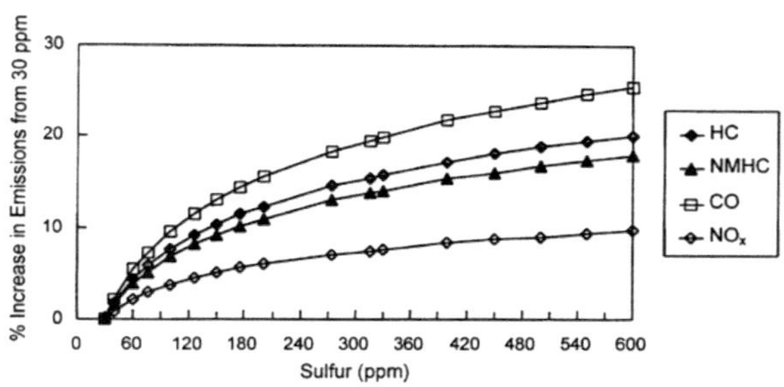

FIGURE 3–12 The percent increase in exhaust emissions for Tier 0 normal-emitting vehicles as a function of fuel sulfur content. The base value for sulfur is 30 ppm. Note that this report usually uses the term VOCs as opposed to HCs or NMHCs to refer to the general class of gaseous organic compounds. Source: EPA 1999r.

The claim that the effects of fuel are reversible is the subject of some controversy and is not reflected in the proposed EPA MOBILE6 sulfur-emissions correction factor. Undoubtedly, additional testing for this effect will be done in the future, particularly on more advanced technology vehicles and vehicles meeting the SFTP. Compliance with the SFTP, beginning with model year 2000, is expected to result in much more precise control of air/fuel ratios that will minimize the opportunities for deposited sulfur in catalysts to be burned off. EPA has estimated that about 50% of the effects of sulfur on catalysts for Tier 2 vehicles can be reversed by reverting to low-sulfur fuel (EPA 1999t). The effects of sulfur on catalysts for Tier 0 and Tier 1 vehicles are fully reversible.

RVP Effects

The Reid vapor pressure is defined as the fuel vapor pressure in pounds per square inch at 100°F. RVP affects vehicle emission primarily in two ways: the fuel evaporation rate and the exhaust emissions. The connec tion between RVP and evaporative emissions is obvious, as one expects a greater evaporative emissions rate from a fuel with a higher vapor pressure. The link between RVP and exhaust emissions is less obvious but well documented. The effects of RVP on exhaust emissions are not significant for NO_x but often are significant for CO and VOCs. Representation of emissions from lower-polluting fuels, especially fuels with lower volatility, was the seventh issue noted in the GAO (1997) report.

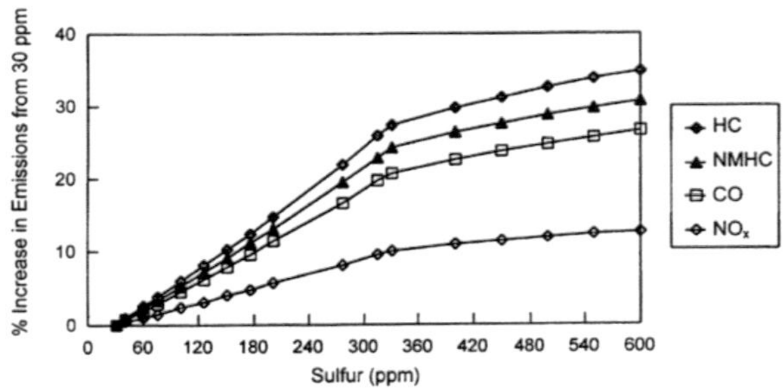

FIGURE 3–13 The percent increase in exhaust emissions for Tier 1 normal-emitting vehicles as a function of sulfur content. The base value for sulfur is 30 ppm. Source: EPA 1999r.

A CRC-AQIRP study (Reuter et al. 1992) found that decreasing the RVP of 9 psi fuels by 1 psi in FTP tests reduced the exhaust CO and VOC emissions by about 4.5% and 9.1%, respectively. An API study (Lax 1994) found that the effect of RVP on exhaust emissions for nominal 10 and 13 psi fuels varied with temperature. The effect was greatest at the highest temperature of the study, 80°F, was reduced at 55°F, and was insignificant at 35°F.

A plausible explanation of the RVP effect on exhaust emissions has been described by Hyde (1998). He suggests that the enhanced CO and VOC emissions occur when a vehicle's evaporative-control system delivers a quantity of fuel to the engine that is so large that enrichment occurs. The evaporative-control system includes a canister packed with activated charcoal to capture excess fuel vapor from the vehicle's gas tank. When the canister becomes loaded, the vapor is released into the engine intake causing the enrichment. The enrichment condition means that the engine is supplied with fuel in excess of the amount that can be completely burned by the available oxygen. The fuel-rich condition and incomplete combustion results in an increase in CO and VOC emissions. High temperatures and high RVP fuels will cause an increase in fuel evaporation, and thus the rate of loading and the frequency of purging fuel from the canister.

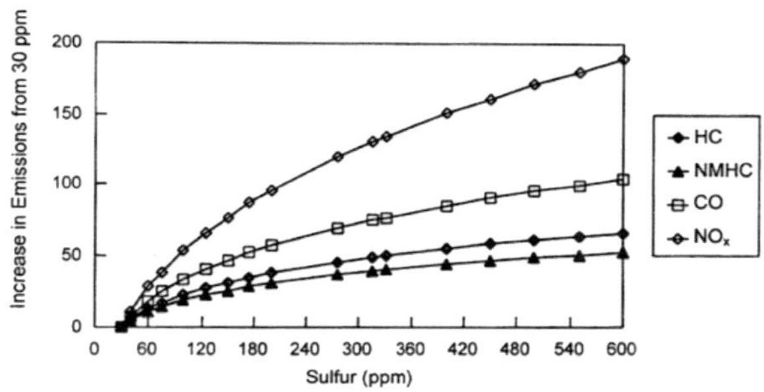

FIGURE 3–14 The percent increase in exhaust emissions for LEV and ultra-low-emitting vehicles (ULEV) as a function of sulfur content. The base value for sulfur is 30 ppm. Source: EPA 1999r.

MOBILE6 exhaust RVP correction factors for RVP are linked to exhaust temperature correction factors (described in the next section). In MOBILE5, RVP correction factors were restricted in that there was no emissions reduction for fuels less than 9 psi (but there was an increase for RVP greater than 9 psi). This was done primarily because there was little test data available at the time for low RVP fuels. At the time of MOBILE5 development, fuels with RVP less than 9 psi were not widely used. The GAO report expressed a concern about this limitation of the model.

For MOBILE6, EPA has not updated the exhaust RVP correction factors, even though there is a wealth of new test data available at low RVP levels, and most areas of the country are using fuels with RVP less than 9 psi in the summer. Although the RVP effects on exhaust emissions are relatively small, credit should still be supplied to states and local areas that are using low-RVP fuels. The exhaust RVP correction factors should be updated in the next version of MOBILE6.

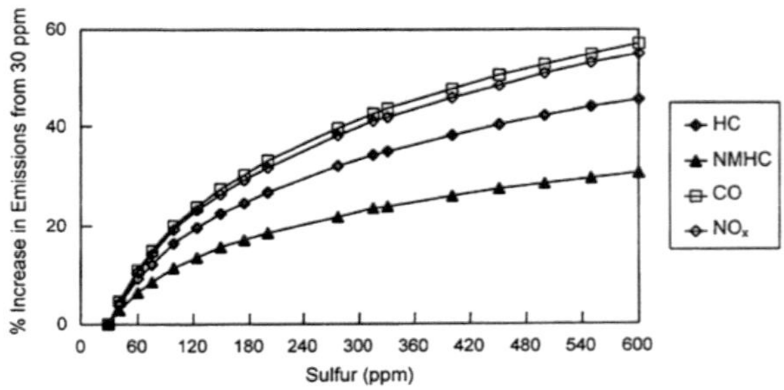

FIGURE 3–15 The percent increase in exhaust emissions for LEV and ULEV trucks as a function of sulfur content. The base value for sulfur is 30 ppm. Source: EPA 1999r.

Hot-soak emissions are estimated as functions of RVP. These functions, derived from test data compiled by EPA, predict "full" soaks in grams per trip. The number of trips per day and fraction of days without trips are factored in to arrive at annual average rates.

Diurnal emissions are handled using the Uncontrolled Diurnal Index (UDI). First, the Wade equation is used to calculate the total diurnal emissions for 9 psi, temperature rise from 60°F to 84°F, and 40% tank level. The UDI is the ratio of the Wade-based emissions of the model's parameter levels to the base conditions. The model then estimates the full, partial, and multiple diurnals as functions of the UDI and finally combines them based upon the relative occurrence of each phenomenon.

Running-loss emissions are calculated using a simple algorithm. Vehicles are stratified by whether they pass a pressure-purge test. For each category, there exist running-loss emissions rates at four temperatures and four RVPs. The model interpolates these to obtain the correct passing and failing running-loss emissions rates and combines these by weighting them according to their expected occurrence in the fleet.

EXHAUST EMISSIONS TEMPERATURE-CORRECTION FACTORS

FTP tests are conducted at a nominal temperature of 75°F using a specified test fuel. Temperature-correction factors (TCFs) are used to ad just the measured FTP emissions to determine the emissions that are obtained at other operating temperatures.

Temperature and RVP corrections are closely related. In MOBILE5, TCFs for temperatures above 75°F are combined with RVP correction factors. Below that temperature, the correction factors are simple multiplicative factors. TCFs are dependent on fuel-delivery technology and are calculated separately for each FTP bag (the FTP and its separate components are discussed earlier in Chapter 3). Technology mixes are then used to combine technology-specific TCFs to produce a TCF for a specific model-year and vehicle-class combination. For Bag 1 CO emissions at low temperature, the correction is in the form of an additive offset with the offset a function of temperature and fuel-delivery type.

EPA (1999u) describes the methods proposed for MOBILE6 for predicting exhaust emissions for vehicle starts and running at non-FTP temperatures. The FTP test is conducted using a dynamometer at 75°F and employs a driving cycle that is intended to be representative of vehicle use in an urban area. Many emissions tests using the FTP driving cycle have been conducted at temperatures above and below the FTP standard of 75°F. These "non-FTP temperature" tests show that in general exhaust emissions of CO and VOCs increase gradually, typically 10% to 30%, with decreasing temperatures from about 80°F down to about 50°F. Below 50°F the emissions increase dramatically in a nonlinear fashion. For example, an API study (Lax 1994) found an increase in VOC emissions of roughly 60% from 55°F to 35°F and an increase in CO emissions of roughly a factor of 2 from 55°F to 35°F.

Because much of the driving in urban areas occurs at non-FTP temperatures, the methods used in MOBILE to predict the non-FTP emissions are critically important to the accuracy of the predictions. The method used in MOBILE6 to predict emissions at non-FTP temperatures involves a linear interpolation between the hot-start (10-min soak) and cold-start (12-hr soak) emissions test points of the FTP driving cycle. The soak time refers to the length of time the hot stabilized test vehicle is turned off and sits at the test temperature before it is restarted and the emission measurement begins. The 12-hour period is the standard for which it has been shown that the test vehicle will come to equilibrium with the ambient temperature of the test chamber. There are two assumptions that appear to be implicit to EPA analysis: (1) the test vehicle changes temperature at a uniform linear rate over a 12-hr period and (2) the emissions levels change uniformly with temperature. Neither of these assumptions is valid. The rate of cooling of a hot engine by convection and conduction is affected by the difference in temperature between the engine and its surroundings. The cooling rate is much faster at first when the engine is hot and becomes slower as the engine cools and its temperature approaches the ambient temperature. The engine's start emissions, which comprise a large fraction of the total driving cycle emissions, are very sensitive to the engine's temperature, so the assumption that the engine cools at a uniform rate over 12 hours is erroneous and arbitrary. As noted above, the effect of temperature on emissions is large and nonlinear below about 50°F. One expects errors in the MOBILE predicted emissions at non-FTP temperatures. The errors from using the linear interpolation method are expected to be largest at lower temperatures, typically below about 50°F.

The MOBILE6 approach to start emissions at FTP temperatures, discussed in the previous section on start emissions, does account for the nonlinear temperature effects. In that approach, the soak temperature is constant at 75°F. The soak function, which depends only on time, accounts for the cooling of the engine and the catalyst as a function of time at the constant ambient temperature of 75°F. A proper approach to temperature corrections for start emissions would use an expanded soak function, which would depend on both ambient temperature and soak time. Such an expanded soak function would require additional data collection efforts.

HEAVY-DUTY VEHICLE EMISSIONS

Heavy-duty vehicles are those exceeding 8,500 lbs gross vehicle weight (GVW). Emissions from these vehicles have come under increased scrutiny recently, particularly those from heavy-duty diesel vehicles (HDDVs). This is due to EPA's past emphasis on control of emissions from LDVs that has increased the relative significance of HDV emissions; public concern over the human health and environmental impacts from PM and NO_x emissions, both of which are emitted in relatively large amounts from HDDVs; advances in emissions-control technologies that have increased the cost-effectiveness of regulating heavy-duty engines; and a recent enforcement action over manufacturers use of "defeat devices" that allows engines to meet emissions standards during certification testing but allows excess emissions during highway driving (EPA 1999v; EPA 1999w). The GAO Report (1997) states the following as one of the major limitations (Issue 11) in the MOBILE Model:

> EPA's supporting data on the in-use emissions of this category of vehicles are about 20 years old; certification standards are higher for these vehicles than their light duty counterparts, and they are generally older and driven more miles annually than their light duty counterparts, although improved emissions technology has lessened the contributions of individual vehicles; some studies are under way, but agency officials

question whether sufficient I & M data will be obtained in time to change emissions estimates for heavy duty vehicles.

The large majority of emissions testing for HDVs has been on an engine dynamometer. There are limited in-use data from I/M programs. Because of these factors, EPA plans to use test data required by EPA from engine manufacturers for new-engine certification as a surrogate for in-use emissions data (EPA 1999v). Engine certification data involve zero-mile level (ZML) emissions plus rates of deterioration to the end of the useful life. Intended service classes and useful lives are shown in Table 3–8. A caveat to the use of these data in developing deterioration rates involves manufacturers observing negative deterioration results. Although testing might indicate engine emissions rates that are lower than the ZML, a manufacturer is not permitted to report a negative deterioration. In those cases, zero deterioration is reported. Therefore, the average deterioration rates calculated from the certification data are higher than the deterioration the manufacturers determined in their laboratory tests. It should be noted that, because all the engines tested for certification receive the proper maintenance and meet the manufacturers' specifications, the effects of inadequate maintenance of the engine and tampering on emissions are not included in the analysis (EPA 1999v).

Another problem using certification data concerns the undercounting of "excess" NO_x emissions. These excess emissions were due to defeat devices used by manufacturers to allow engines to meet emissions levels during testing, but switches off emissions controls for improved fuel economy during highway operations. These defeat devices involve engines since 1988 having software that advances the injection timing under high-speed operating conditions and increases the NO_x emissions above the FTP transient cycle certification levels. Recently, EPA has entered into a consent decree

TABLE 3–8 Intended Service Classes and Useful Lives for Heavy-Duty Engines

Engine Class	Gross Vehicle Weight (GVW) (lb)	Useful Life (miles)
All heavy-duty gasoline engines	,501–60,000+	110,000
Light heavy-duty diesel engines	8,501–19,500	110,000
Medium heavy-duty diesel engines	19,501–33,000	185,000
Heavy heavy-duty diesel engines (incl. buses)	33,001–60,000+	290,000

Source: EPA 1999v.

with, the diesel engine manufacturers requiring manufacturers to offer new software when 1988–1998 engines are rebuilt. These effects will be incorporated into MOBILE6 (EPA 1999w). Further field data regarding the effect of the rebuild kits will need to be gathered for use in improving the modeling of NO_x emissions in a revision of MOBILE6.
Source: EPA 1999v.

Although little is known to date about HDV high emitters, there is now starting to be some state effort to smoke test vehicles on the road. This will not likely give data about the regulated emissions of HC, CO, NO_x and PM but will give some data about the state of maintenance of vehicles on the road. There is a need to determine data on high emitters, and the state programs could provide a source of vehicles that have high smoke emissions, which could then be tested for other emissions.

Because of the lack of I/M data, MOBILE6 will not have accurate emissions rates that represent in-use emissions from HDVs. This is a serious shortcoming of the MOBILE6 model since the technology and the emissions levels have changed during the past 25 years. However, extensive data are now being gathered in a number of major research programs. These data should be analyzed and interpreted relative to determining better deterioration factors and improved in-use emissions factors from HDVs.

Use of Engine Data and Conversion Factors To Estimate Vehicle Emissions

For MOBILE6, EPA (1998i) has updated the estimates of heavy-duty engine emissions factors currently contained in MOBILE5b. The methodology involves the estimation of a gram per mile emissions factor by multiplying a work-specific emissions level (in g/bhp-hr) by a conversion factor that converts work units into mileage units (bhp-hr/mi). In mathematical form, the conversion factors can be expressed as

$$\text{Conversion Factor (bhp - hr/mi)} = \frac{\text{Fuel Density (lb/gal)}}{\text{BSFAC (lb/bhp - hr)} \times \text{Fuel Economy (mi/gal)}} \tag{3–6}$$

where BSFC is the brake-specific fuel consumption.

EPA divides heavy-duty vehicles into several classes in MOBILE. Table 3–9 shows that the model classifies heavy-duty vehicles into those using gasoline or diesel fuels. These are further subdivided based on GVW. EPA uses the classifications in Table 3–9 to account for different characteristics and general uses of the engines included in each GVW class.

TABLE 3–9 Heavy-Duty Vehicle Classifications for MOBILE6

Designation	Description	Gross Vehicle Weight (lb)
Gasoline Vehicles		
HDGV (classes 2B-3)	Heavy-duty gasoline vehicles	8,501–14,000
HDGV (classes 4–8)	Heavy-duty gasoline vehicles	>14,000
Diesel Vehicles		
HDDV (class 2B)	Light heavy-duty diesel trucks	8,501–10,000
HDDV (class 3)	Light heavy-duty diesel trucks	10,001–14,000
HDDV (classes 4–5)	Light heavy-duty diesel trucks	14,001–19,500
HDDV (classes 6–7)	Medium heavy-duty diesel trucks	19,501–33,000
HDDV (class 8A)	Heavy heavy-duty diesel trucks	33,001–60,000
HDDV (class 8B)	Heavy heavy-duty diesel trucks	>60,000
Urban Buses		
HDGB (school)	Heavy-duty gasoline school buses	All
HDGB (transit)	Heavy-duty gasoline transit buses	All
HDDB (school)	Heavy-duty diesel school buses	All
HDDB (transit)	Heavy-duty diesel transit buses	All

Source: EPA 1999x.

EPA decided to recompute the emissions levels and deterioration rates based on model-year groups that represent changes in EPA's emissions standards. To improve the flexibility of MOBILE6's emissions factors, EPA has chosen to use emissions rates for each service class as shown in Table 3–9 instead of a single rate as used in MOBILE5. For heavy-duty gasoline engines, EPA will continue to have a total emissions rate in MOBILE6. These changes in MOBILE6 represent improvements over MOBILE5.

However, all of the different classes of vehicles shown in Table 3–9 likely have different driving cycles. At present this is not accounted for completely in MOBILE5 (and for that matter, in MOBILE6), because the in-use emissions in grams per mile is based on FTP transient dynamometer emissions data multiplied by a conversion factor that accounts mainly for fuel economy differences and VMT as a function of vehicle age. There is a need to develop more modal emissions data and driving-cycle data by vehicle class so improved emissions in grams per mile can be made available in a revision of MOBILE6.

Adjustments to Heavy-Duty Vehicles Basic Emissions Rates

The MOBILE6 model will calculate emissions factors at low and high altitudes. Low-altitude emissions factors are based on conditions representative of approximately 500 feet above mean sea level and high-altitude emissions factors represent conditions of approximately 5,500 feet above sea level. At the present time, EPA has been unable to obtain recent studies as to the effects of varying altitude on exhaust emissions from heavy-duty gasoline vehicles. Therefore, MOBILE6 will use the same adjustment factors used in MOBILE5. EPA was able to locate a small number of heavy-duty diesel vehicle studies evaluating the effects of altitude on changes in emissions. From these data, EPA developed new altitude-adjustment factors for heavy-duty diesel vehicles in MOBILE6 (EPA 1999v).

At present, MOBILE5 does not account for the effects of temperature or humidity on HDDV emissions. Temperature and humidity vary in different parts of the United States and they will very likely affect NO_x and also PM emissions. The Engine Manufacturers Association (EMA) is presently conducting tests at Southwest Research Institute (SwRI) to answer this question. EPA should review these data when they become available in late 1999 and incorporate correlations into a revision of MOBILE6. Recent tunnel data may show that HDVs emit more NO_x then predicted by MOBILE5 (J.C.Sagebiel, Desert Research Institute, personal commun., May 1999). One of the explanations for this difference could be temperature and humidity effects, along with the fact that the FTP NO_x data might not be representative of in-use NO_x.

Finally, diesel fuel properties have been shown to have an effect on emissions. The properties of cetane number, aromatics, and sulfur content have been most often cited as the important variables. Sulfur content is presently in PART5 relative to SO_2 and sulfate emissions. EPA should continue to evaluate the available data to see if reasonable approximate correlations of these variables can be developed for MOBILE to relate fuel properties to the NO_x and PM emissions from HDDV.

PARTICULATE EMISSIONS

As noted in the introduction to this chapter, PM emissions from on-road mobile-sources are estimated using the PART5 model. PART5 is a separate Fortran program, but is compatible with MOBILE5a in format and fleet characterization. PART5 calculates exhaust PM as the sum of lead (for gasoline-powered vehicles), soluble organic fraction (SOF), remaining carbon portion (RCP), and directly emitted sulfate (SO_4). The emissions from tire and brake wear are also calculated in the model.

PART5 calculates PM factors in grams per mile for 12 vehicle classes: light-duty gas vehicles, light-duty gas truck 1, light-duty gas truck 2, heavy-duty gas vehicles, motorcycle, light-duty diesel vehicle, 2B heavy-duty diesel vehicles, light heavy-duty diesel vehicles, medium heavy-duty diesel vehicles, heavy heavy-duty diesel vehicles, and buses. It reports emissions as a function of particle sizes less than or equal to 1.0 to 10.0 μm. The fraction in each size range is hard-coded in the model. For example, PART5 assumes 97% of exhaust particulate is 10 mm or less for unleaded gasoline-fueled vehicles.

Although the documentation covering PART5 (EPA 1995) does not explicitly address the source of the underlying data, communication with EPA staff confirm that most of the estimates for exhaust emission of HDVs and LDVs are based on either certification data and analyses dating back to early 1980, or on ratios to hydrocarbon emission factors. The woefully outdated emission factor estimates in PART5 make the accuracy of the model highly suspect; the emission factors must be updated using results of recent test programs for both light-duty and heavy-duty vehicles (discussed below).

Exhaust Particulate Emissions

Exhaust PM have typically been a concern for diesel-powered vehicles, which are predominately used in heavy-duty vehicles. Concurrent with EPA's consideration of an ambient air-quality standard for particle sizes of 2.5 mm and less, there has been increased concern that gasoline-powered engines, the main propulsion source for the light-duty fleet, could make a significant contribution to ambient PM. Recent studies of PM emissions from LDVs, such as the CRC Projects E-24–1 (Cadle et al. 1998), E-24–2 (Norbeck et al. 1998), and E-46 (Cadle et al.1999), as well as the Northern Front Range Air-quality Study (Norton et al. 1998), show PART5 emissions rates to be lower than those observed and that a significant fraction of emissions are coming from LDVs.

Tables 3–10 and 3–11 show this underestimation of PART5 emissions rates for a range of LDVs. Table 3–10 contrasts the results of a test performed for CRC Project E-24–2 with estimates produced by PART5 for light-duty passenger cars and trucks. This underestimation of exhaust PM emissions in PART5 is also evident in the comparison of PART5 rates with the weighted FTP results from the CRC E-24–1 and E-46 projects. Table 3–11 displays those results.

Although more attention has been given to particulate measurements from heavy-duty diesel engines, PART5 did not have sufficient data to determine a deterioration rate for PM emissions for these vehicles. A recent report prepared for EPA (Weaver et al. 1998) shows significant deterioration for 1994 and later trucks and transit buses. In addition, work by Graboski et al. (1998) concluded that PART5 was "significantly underestimating" PM emissions from heavy-duty vehicles.

TABLE 3–10 PM Emissions Rates in g/mi from PART5 and from the CRC Project E-24–2

	Passenger Cars		Light-Duty Trucks	
Model Year	CRC	PART5	CRC	PART5
Pre-1981	.315	.193	.368	.193
1981–1985	.486	.025	.493	.025
1986–1990	.172	.006	.122	.006
1991–1997	.018	.004	.035	.004

Source: Cadle et al. 1998.

It is important that EPA revise PART5 to reflect in-use PM emissions. This will likely require extensive field measurements. Data are also needed to assess differences between HDDV PM measurements obtained in the laboratory compared to in-use emissions; EPA should review work currently being conducted by the CRC and the National Cooperative Highway Research Program in this area. In addition, data are needed relative to systems of the engine and the vehicle that reflect maintenance problems that affect emissions. Studies are also needed on the effectiveness of diesel I/M programs and whether smoke I/M programs tend to increase NO_x

TABLE 3–11 Passenger Car PM Emissions Rates in g/mi from PART5 and from the CRC Projects E-24–1 and CRC E-46

Model Year	CRC E-24–1	PART5
Pre-1981	.955	.193
1981–1985	.474	.025
1986–1990	.444	.006
1991–1997	.028	.004
	CRC E-46	PART5
Tier 0	.083	.025
Tier 1	.038	.006

Source: Norbeck et al. 1998; Cadle et al. 1999.

emissions and reduce PM emissions. Ways of screening and characterizing the in-use vehicle population for high-emitting vehicles are needed. EPA needs to determine the effectiveness of the catalytic oxidation devices used since 1994 to reduce PM emissions by reducing the SOF. There is no information on the long-term effectiveness, maintenance practices, or tampering for these devices.

Source: Norbeck et al. 1998; Cadle et al. 1999.

Particulate Emissions from Tire and Brake-Wear

PART5 estimates emissions from tire wear based on the assumption that the emissions rate of airborne particulate is 0.002 grams per mile per wheel (EPA 1998j). This reference, known as AP-42, lists two studies as the basis of the estimate for tire-wear emissions (Williams and Cadle 1978; Brachaczek and Pierson 1974). The single emissions rate is based on tests of LDVs and no estimate for the airborne particle size distribution for tire-wear is offered. Tire-wear emissions less than 10 μm are based on interpolation. The dated references for the PART5 emissions factors suggests that these factors are based on tests of older biased-ply tires rather than longer-wearing tire technologies currently in use.

PART5 reports brake wear as a separate emissions factor of 0.0128 grams per mile, based on a paper by Cha et al. (1983). Brake-wear particulate emissions are higher than for tire wear because a larger fraction is assumed to be less than 10 μm in diameter. Brake-wear emissions factors in PART5 are assumed to be the same for all vehicle classes, although it could be assumed, as with tire wear, that the number of wheels, the weight of the vehicle, and the driving cycle would be significant contributing factors related to the per mile emissions rate. As with tire wear, the dated reference suggests that the emissions factor is based on older materials and needs to be updated.

Issues for Model Revision

EPA is planning to update PART5 after it completes the updates for MOBILE6. As recommended above, such a revision should become part of MOBILE rather than being issued as a separate model. Issues that need to be addressed while updating PART5 are listed below:

- Data on particulate emissions from HDVs need to be updated to include the effects of deterioration in emissions, adjustment for the benefits of I/M, and variations due to actual driving conditions.

- Data from new studies on emissions from light-duty gasoline-powered and diesel-powered vehicles should be included in the updated particulate emissions model. Data collection efforts should be expanded to ensure that the effects of deterioration, I/M, and off-cycle driving conditions are included in the model.
- Data on modern tires and brake materials must be obtained for inclusion in future particulate emissions inventories.

FLEET CHARACTERIZATION

In MOBILE, the fleet is characterized by three parameters: age or registration distribution, mileage-accumulation rate, and fleet or VMT mix. The age distribution gives the fraction of all vehicles in a particular class that are of a certain age. Because MOBILE5 accounts for 25 different ages (except for motorcycles), 25 fractions are required for each vehicle class, with the fractions summing to unity within each class. The mileage accumulation rate is the annual number of miles a vehicle is expected to be driven. It varies by vehicle age and class. The fleet mix gives the fraction of the fleet total VMT traveled by each of the eight vehicle classes. Again these fractions must sum to one.

MOBILE calculates a vehicle class's emissions factor by computing the emissions factors for each of the model years, weights these by each model year's contribution to the vehicle class's total annual VMT, and then sums the weighted emissions factors. The weighting factors are termed travel fractions. The travel fraction, TF_m, represents the fraction of the total VMT that is accounted for by a vehicle of age m years. It is calculated from the fraction of vehicles registered that are m years old, REG_m, and the annual mileage accumulation for these vehicles, $MILES_m$,

$$TF_m = \frac{REG_m * MILES_m}{\sum_{k=1}^{MaxYears}(REG_k * MILES_k)}$$

(3–7)

where the summation is over all model years k. Once the vehicle-class-specific emissions factors are computed, the model then weights each of them by the corresponding fleet mix fraction and sums the results to produce the fleet emissions factor.

Although the user is allowed to enter custom registration distributions, mileage-accumulation rates, and fleet mixes, the model contains default (or for fleet mix, internally calculates) values for these parameters. The default registration distributions are obtained from sales fractions through the 1980s. Mileage accumulation rates in MOBILE5 are based on 1990 National Purchase Diary (NPD) data. Fleet mix is internally calculated using registration distributions, mileage accumulation rates, diesel sales fractions, and total vehicle counts by class.

MOBILE6 will see changes in all of the above parameters. These changes are documented in an EPA report (1999x). They are brought about by the availability of new data as well as shifts in methodology in some cases. The new model expands the previous 8 vehicle classes to 28, thus requiring much more detailed fleet characterization data. New mileage-accumulation rates for light-duty vehicles and light- and heavy-duty trucks are derived from the 1995 National Personal Travel Survey (NPTS) and the 1992 Truck Inventory and Use Survey (TIUS), respectively. Revised registration distributions are obtained from 1996 data compiled by the R.L.Polk company. Vehicle counts are based on data from various sources including the 1996 Polk data, 1998 Certification and Fuel Economy Information System (CFEIS) database, the *Annual Energy Outlook* (Energy Information Administration 1998), and a report by Ward's Communications (Pemberton 1996) which gives scrappage rates.

Figures 3–16, 3–17, and 3–18 give sample comparisons of the MOBILE5 and MOBILE6 mileage-accumulation rates, registration distributions, and vehicle counts. Note that the new registration distribution is smoothed. This is a significant change from the former approach, which reflected actual historical trends in sales and perpetuated them in all future-year calculations.

SUMMARY AND RECOMMENDATIONS

EPA is currently developing MOBILE6, the newest version of the MOBILE model. This version is scheduled for release in the year 2000. In developing MOBILE6, EPA has used a more-open process, involving many stakeholders, and has been published much documentation for general review.

In the development of MOBILE6, EPA has addressed many shortcomings of MOBILE5, particularly those identified in the GAO (1997) report. The extent to which MOBILE6 has addressed those concerns is summarized in Table 3–12. The table also notes some improvements that can be made in future versions of MOBILE.

In addition to the recommendations in Table 3–12, the committee offers the following recommendations for the improvement of MOBILE. These recommendations begin with changes to components in the existing

MOBILE model. The next set of recommendations pertains to models that are closely related to MOBILE, such as those that estimate PM and air toxic emissions. The final recommendation deals with the need for long-range planning to guide the future development of the model.

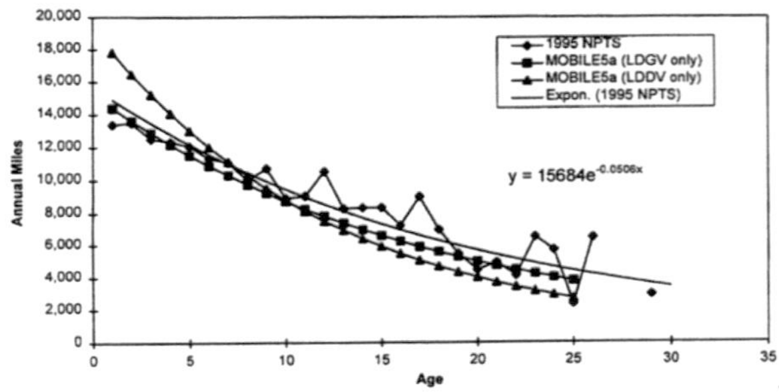

FIGURE 3–16 Light-duty vehicle annual mileage-accumulation rates. MOBILE6 uses the curve developed from the 1995 NPTS data. Source: EPA 1999x.

Obtain Better Data on High-Emitting Vehicles

Establish a long-term testing program to characterize in-use deterioration of representatively aged new-vehicle technology using a driving cycle more representative of actual driving conditions. This should focus on determining the nature of both exhaust and evaporative high emitters. Improved data on both the emissions rate and the fraction of the vehicle population that are high emitters are required.

Inclusion of Road-Grade Effects in MOBILE

The emissions increase from road grade is similar to that from acceleration and should be included in the model. This will be particularly impor tant for areas where there is a significant amount of grade such as Denver and Spokane. Planning for this feature should include input from local regions that use MOBILE to ensure that grade information is available to potential users and that the model revisions are consistent with the available formats of the grade data.

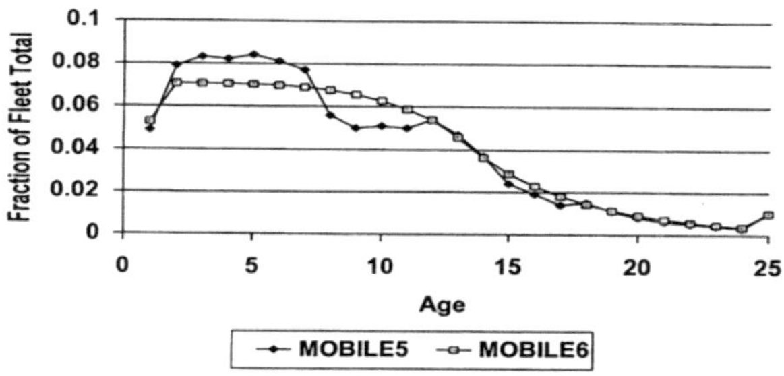

FIGURE 3–17 Comparison of MOBILE5 and MOBILE6 light-duty vehicle registration distribution. Source: EPA 1999x.

Improve the Start-Emissions Database

Routine tests of start emissions should be made as part of ongoing measurement programs unless there is confirmation that regression techniques, similar to those used for MOBILE6, provide an effective estimation of start emissions. Additional measurements of the effects of ambient temperature, wind speed, and soak time on start emissions should be made to get a better representation of these important factors. Another factor that should be considered when estimating start emissions is the operating mode of a vehicle during the first minutes of operation.

Modeling of Inspection and Maintenance Programs Benefits

In particular, the treatment of vehicles that failed emission tests but never appeared for a retest, owners who never have their vehicles in spected, and the deterioration of vehicles after repair needs to be improved. The modeling of repaired vehicles' deterioration should be based on data from actual repaired vehicles.

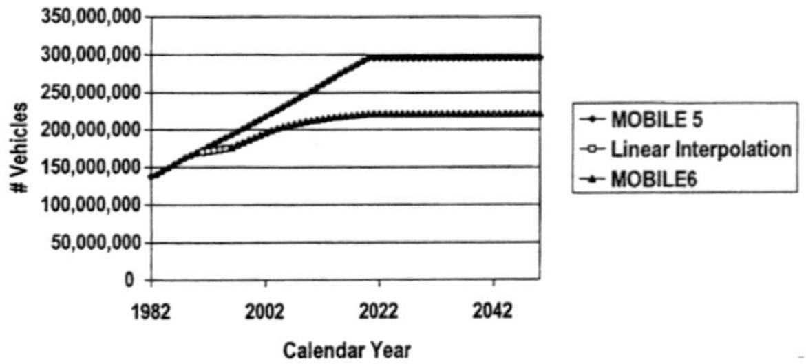

FIGURE 3–18 Comparison of light-duty vehicle counts, 1982–2050. Source: EPA 1999x.

Improve the Emissions Factors for Heavy-Duty Vehicles

Emissions factors for HDVs are woefully outdated and there are questions about the conversion of engine dynamometer data into on-road gram-per-mile emissions. Appropriate chassis dynamometer cycles need to be developed for HDVs and data must be obtained on such cycles. Appropriate corrections for the effects of humidity and temperature, currently under development, should be incorporated into MOBILE. Data should to be generated for in-use conditions that might have significantly different emissions from those predicted based on engine certification tests.

Updating of Fleet Characterization

In recent years, there has been a significant increase in the use of light-duty trucks (especially sport-utility vehicles) instead of and in addition to passenger cars. EPA has updated fleet characterization data for MOBILE6 to reflect these current trends. EPA should at regular intervals (every 2 years or so) review the fleet characterization data, both current and pro

TABLE 3–12 Summary of Expected Changes to MOBILE6 That Respond to Problems Identified in GAO Report (1997)

Area of Concern Regarding MOBILE Model Cited in GAO Report	MOBILE6 Treatment of Issue
1. Emissions estimates for higher speeds, especially speeds in excess of 65 mph. 2. Representation of emissions from rapid acceleration and deceleration, including aggressive driving behaviors.	MOBILE6 uses data obtained from recent studies on real-world driving conditions to develop facility-specific speed-correction cycles, which include higher speeds and aggressive driving behavior. The facility-specific speed-correction factors also provide greater distinction in roadway classifications.
3. Representation of emissions immediately after engine start-up, known as cold-start emissions.	Start emissions have an improved treatment in MOBILE6; more study should be done to provide additional data for the approach proposed.
4. Representation of emissions from air conditioner use.	MOBILE6 has an improved model of air-conditioner use. Additional data and model modifications could improve the estimates of this effect.
5. Representation of emissions from road grade, such as when a car climbs a hill.	Not addressed in MOBILE6.
6. Representation of high-emitting vehicles in the MOBILE's supporting database.	EPA used data from IM240 lanes to correct FTP data for recruitment bias in exhaust-emissions data. Special studies should be done to determine effect of high emitters.
7. Representation of exhaust emissions from lower-polluting fuels, especially fuels with lower volatility (low RVP); representation of emissions from oxygenated fuels.	MOBILE6 exhaust emissions effects of low RVP fuels has not changed from MOBILE5. MOBILE6 shows reduced benefits from oxygenated fuels, based on EPA analysis of more recent test data.

jected, to ensure that changes in the vehicle fleet are properly recognized in the emissions model.

Area of Concern Regarding MOBILE Model Cited in GAO Report		MOBILE6 Treatment of Issue
8.	Representation of emissions system deterioration for vehicles with 50,000 or more miles.	New data have shown much lower emissions rates for such vehicles. These data have been used in MOBILE6.
9.	Emissions estimates and assumptions for vehicle I/M programs.	MOBILE6 shows reduced benefits from I/M programs. Questions remain about the assumed benefits for OBD and the assumed deterioration of repaired vehicles.
10.	Estimates and assumptions for nontailpipe evaporative emissions when the vehicle is not operating.	MOBILE6 includes updates to resting loss emissions based on real-time (24-hr) test data.
11.	Emissions estimates and assumptions for the inspection and maintenance of HDVs—those with a gross vehicle weight of 8,501 pounds or more.	Not included in MOBILE6.
12.	Data characterizing vehicle fleet.	EPA has updated fleet characteristics (fleet mix and age and mileage-accumulation distributions by vehicle class) for MOBILE6.
13.	Greater distinctions in roadway classifications.	See response to items one and two.
14.	Quantifying the uncertainty of the model's estimates.	Not included in MOBILE6.

Complete Documentation of all Databases and Analyses

EPA has done an excellent job of improving their documentation of the basic steps in the MOBILE model. Additional documentation should be provided to explain all the details of the analyses so that interested parties can readily check the analyses. Placing these detailed analyses and *all* the databases used on the internet would facilitate external review of the databases and methods used in MOBILE.

Integration of the PART5 and MOBILE Models

PART5 and MOBILE do basically the same thing: compute actual emissions from on-road mobile sources. The separation of gaseous emissions in MOBILE and participate emissions in PART5 is not necessary. It requires users to run two models instead of one and leads to the possibility that the on-road motor vehicle fleet and other important factors will not be treated consistently between the two models.

When incorporating the PART5 model into MOBILE, several problems with PART5 need to be addressed. Updated emissions factors should be developed incorporating data on the effects of high-emitting vehicles (smoking vehicles), in-use deterioration, I/M programs, speed variations, and off-cycle emissions.

Incorporation of the COMPLEX Model into MOBILE

The effects of reformulated gasolines are currently estimated in a separate model called the COMPLEX model. Incorporating a COMPLEX-like model into MOBILE would allow states and regions to directly model gasoline formulations with more stringent requirements than federal reformulated gasoline requirements. It would require the impacts of reformulated gasolines to be extended to include impacts on emissions from all vehicles and technology groups as well as the impacts on CO emissions.

Incorporation of Toxic-Emissions Factors into MOBILE

These emission factors are currently in a separate model, MOBTOX. The rationale for this recommendation is the same as that for the two previous recommendations: convenience and consistency.

More Emphasis on the NONROAD Model

Although the EPA model, NONROAD, for off-road emissions sources is not part of MOBILE, the committee notes that this critical emissions model is lacking in data on emissions factors and activity levels. As more controls are placed on on-road sources, the off-road sources will become more important in the future, and the NONROAD model will play a larger role in estimating regional emissions inventories.

Develop Long-Range Plans for the Evolution of the MOBILE Model

The future implementation of new emissions and fuel standards, growing concerns about PM and air toxics emissions, and the rising cost of control strategies will increase the focus on MOBILE. Users will need improved accuracy and reliability from MOBILE, even as the regulatory setting, vehicle technologies, and fleet characteristics are changing. This poses a daunting challenge for EPA's Office of Transportation and Air Quality (OTAQ) to develop an accurate model that reflects uncertain regulations, unproven technologies, and shifting preferences among consumers.

To address these demands, EPA must develop a long-range plan for addressing critical modeling issues. EPA first should determine the most appropriate uses for the model and develop improvements to support these specific uses. This plan should then set priorities for model improvements that have the largest impact on emissions and develop a plan for collecting the necessary data to support these improvements. Most importantly, EPA must also address how close the modeling of mobile-source emissions should be to the development of regulations. The model must be seen as an accurate reflection of mobile-source emissions, not as a tool that is used to support proposed regulations.

4

Model Uncertainty and Evaluation

IS A TOOL for estimating current and forecasting future mobile-source emissions, including quantifying the effects of control measures. These results form key elements of many air-quality regulatory compliance programs and directly affect transportation planning and the selection of control strategies. Thus, there is a need for a high degree of accuracy from MOBILE.

One of the specific charges to this committee is to assess, to the extent practical, the overall accuracy of the current version of the MOBILE model in predicting atmospheric emissions. Such information is derived from model evaluation. However, the necessary information to quantitatively evaluate the current or upcoming versions of MOBILE with great confidence does not exist. This lack of information is one of the most serious concerns with MOBILE and its use.

Assessing accuracy in the context of MOBILE involves model evaluation, typically by comparing real-world emissions with model predictions. It also involves considering uncertainty and bias that arise from the wide variety of observations, assumptions, and mathematical relationships that underlie MOBILE model algorithms. This chapter discusses uncertainty and model evaluation[1] and reviews previous studies on these topics.

[1]This report uses model evaluation in reference to assessing the ability of a model to accurately represent the real world, for example by being able to estimate the emissions from mobile sources with little error. This differs from model validation that refers to assessing the correctness of the form of the model. MOBILE is based on statistical analyses of data, and the form of MOBILE is a set of statistical relationships. Thus, model validation would correspond to determining whether the statistical relationships were derived in a valid manner and implemented correctly. Model evaluation determines whether those relationships provide accurate results.

DEFINITION OF TERMS

Accuracy, Uncertainty, and Bias

Accuracy in MOBILE refers to its ability to correctly estimate the true value of emissions. Bias is the tendency for estimates to be consistently higher or lower than the true values. Uncertainty is the variability (or scatter) in MOBILE's prediction about the actual emissions. MOBILE is accurate, for example, if it predicts the correct emissions factors for the fleet of on-road vehicles, and if it predicts the actual changes that result from mobile-source emissions control programs such as inspection and maintenance (I/M). (MOBILE, as typically applied, provides only point estimates without any statistical confidence intervals around those estimates). MOBILE predictions can be assessed by comparing accurate in-use measurements of vehicle emissions and air quality to model predictions. This is termed model evaluation and is the subject of the second half of this chapter.

Uncertainty and bias in MOBILE arise from many sources, primarily from the data used to construct the model, and from errors in analyses and assumptions leading to model formulations (discussed further below). Uncertainty and bias in MOBILE are difficult to assess because of the complexity of the model, the uncertainty in the underlying emissions data and model formulation, the uncertainty in the input data, and the difficulty in obtaining accurate measures of real-world emissions (e.g., from analyzing ambient data).

Uncertainty, as used here, should not be confused with repeatability. If MOBILE is run several times with the same set of inputs, the model will always generate exactly the same output. The model is repeatable simply because it has no stochastic or random component; this does not by any means imply that the model is accurate.

Figure 4–1 shows the difference between bias and uncertainty. The top box in Figure 4–1 shows a case in which the model provides accurate results. The bottom box shows estimates that are uncertain (in that they are scattered about a one-to-one correspondence) but unbiased. The middle box shows estimates that are biased; *on average* the predictions are below the actual emissions. For MOBILE's many uses, it is important that the model's estimates be accurate with a high degree of certainty and

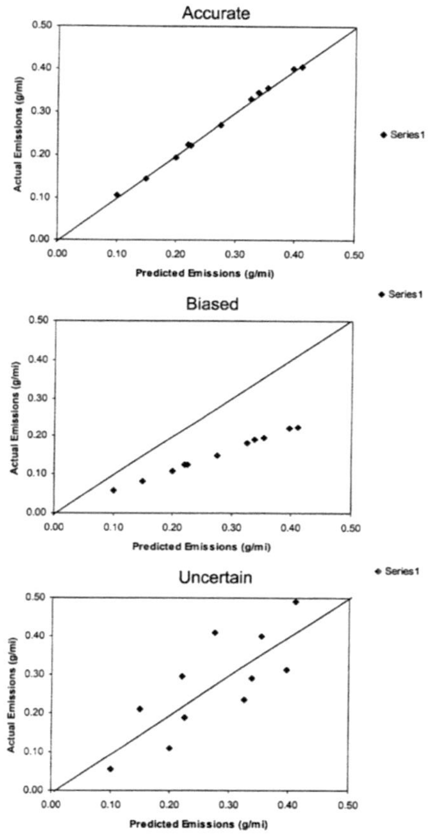

FIGURE 4–1 Representation of bias and uncertainty (hypothetical data).

a low bias. If not, transportation and air-quality planners could be led to implement costly, unnecessary control programs.

Sensitivity

Model sensitivity refers to the variation in model output in response to changes in model inputs such as average speed, ambient temperature, fuel volatility, and I/M program parameters. It is important that air-quality planners understand the effects of changes in these model inputs. The U.S. Environmental Protection Agency (EPA) has provided sensitivity analyses for some earlier versions of MOBILE (see discussion below), but not extensive analyses for MOBILE5. A comprehensive sensitivity analyses should be performed for *all* model inputs and provided as part of user guides for all future versions of MOBILE.

TYPES AND SOURCES OF UNCERTAINTY AND ERROR

Many kinds of uncertainty plague the emissions estimates provided by MOBILE. A major fraction of those uncertainties arise from limitations in the scientific and technical basis of MOBILE as well as the data inputs, including sampling and measurement errors. As described above, the primary source of uncertainty in MOBILE output is in the underlying emissions data used to generate the model formulations. There can also be limitations in MOBILE's structure. Input uncertainties, including data on vehicle characteristics and usage, are propagated through the model and contribute substantially to uncertainty. Another source of uncertainty that should be analyzed is the true variability in the system being modeled.

It is clear that it is not practical to eliminate all uncertainty from mobile-source emissions models. MOBILE is a statistically based model, and its accuracy depends on valid and comprehensive samples as the foundation of the statistical relationships within. Some uncertainty comes from the random variation in the relatively small samples. No sample short of 100% (a complete census) can be large enough to completely eliminate randomness from affecting parameter error, but in all practical circumstances, the influence of random variation on an estimate of a parameter decreases as the sample size gets larger. Even if 100% of the on-road vehicles were tested, uncertainty would still arise from the incompleteness and unrepresentativeness of the tests and errors in the testing procedure.

Analysts must also consider true variability and its impact on uncertainty. Within defined fleets, manufacturers do not produce exactly the same kinds of vehicles. Vehicles from a single manufacturer are not iden tical. Also, emissions from any one vehicle vary from time to time and place to place in ways that are correlated with age, mode of driving, and many other things. Fleet characteristics vary spatially, as do topography and driving habits. The contribution of this true variability to the overall uncertainty of MOBILE is unknown, but could be partially characterized in an advanced mobile-source emissions model if the data (both input and model formulation) were available.

Uncertainties further arise from limitations inherent in the model's structure. For example, no fleet of vehicles is well characterized. Roadway networks and associated driving patterns are not perfectly represented. MOBILE does not capture all the factors leading to emissions (particularly high emissions, such as those induced by steep road grades). These and other model limitations discussed in Chapter 3 influence uncertainty.

Following are descriptions of the general types of uncertainties that arise in the MOBILE model. Several examples of each type are provided, for both MOBILE5 and MOBILE6. For additional information, see the study by Wenzel et al. (In press), which discusses emissions variability in more detail, and also describes other issues that complicate the statistical analysis of vehicle-emissions test data.

Nonrepresentative Vehicle Samples

MOBILE algorithms and emissions factors are based largely on test data from in-use vehicles that are solicited through the mail or by recruitment at I/M test stations. Typically, the owner is provided with a small payment and a rental vehicle until testing is completed. Free vehicle repairs are sometimes provided as an incentive. Response rates for such recruitment efforts are very low, typically less than 25% and sometimes as low as 5%.

Recruited vehicles have serious bias issues because high emitters and tampered vehicles as well as expensive luxury vehicles are less likely to be voluntarily submitted for testing. As discussed in Chapter 3, very high emitting vehicles are a relatively small fraction of the on-road vehicle fleet, but they contribute a very large fraction of total vehicle emissions. Emissions from high-emitting vehicles are much more variable than emissions from normal emitters (Knepper et al. 1993), and thus require a large sampling fraction to obtain reasonably accurate estimates of their emissions (and to estimate the effects of control programs for their emissions). It is thus critically important that such vehicles be appropriately represented in emissions testing programs. If emissions from high emitters are not properly characterized, then MOBILE emissions factors can be seriously biased.

EPA recognizes that underrepresentation of high-emitting vehicles in the Federal Test Procedure (FTP) and IM240 databases underlying MOBILE5's basic emissions rates is a serious shortcoming that produces biased (low) emissions estimates. For MOBILE6, EPA is proposing to adjust the basic emissions rates based on data from the Dayton, Ohio, IM240 program. Although this is a step in the right direction, there are still biases in the Dayton IM240 data. One type of bias arises from noncompliance with the program—the database does not include emissions from vehicles that are registered in the I/M area but do not obtain the required inspection (and repair if needed). Remote sensing studies have shown that these noncomplying vehicles have higher than average emissions (e.g., Stedman et al. 1997; Stedman et al. 1998; Wenzel 1999). In fact, the I/M program compliance rate is one of the inputs to MOBILE, and the model assumes that noncomplying vehicles have emissions about twice as high as complying vehicles. A second problem is that when the Dayton program was first implemented, there is strong evidence that owners of vehicles that had been registered in the I/M area changed their registration to surrounding counties that were not subject to the Dayton I/M program (McClintock 1999). Again, these vehicles are more likely to be higher emitting vehicles. Although EPA recognizes the inherent biases in their testing data, use of the Dayton IM240 data to adjust for those biases still likely results in biased emissions estimates.

Some of the MOBILE algorithms and correction factors are based on very small samples, which may be nonrepresentative of the population. An example for MOBILE6 is the test data used to derive air-conditioning correction factors for light-duty vehicles (LDVs). As described in Chapter 3, the effects of full air-conditioning operation on vehicle emissions were determined from a sample (of 37 vehicles)—23 passenger cars and 14 light-duty trucks, all from model years 1990 to 1996. Only five of the vehicles were high emitters for at least one pollutant. Similarly, the activity data for air-conditioning operation were obtained from a fleet of only 20 vehicles operating in Phoenix, Arizona, from August to October 1994. And, although the actual air-conditioning operation depends on the torque generated by the compressor, no data were available for this variable.

Variability in Vehicle Emissions

Vehicle emissions are highly variable, for a variety of reasons. Two vehicles of the same make and manufacturer, model year, technology, and accumulated mileage can have very different emissions measured on the same test or drive cycle. Such variation can be caused by factors such as how the vehicle has been driven and maintained, prior tampering with emissions control system components, and repeated excessive driving loads. Studies have also shown relationships between socioeconomic factors and vehicle emissions, with vehicles in lower than median income households exhibiting higher than average emissions (Singer and Harley 2000).

Vehicles of the same age and technology also exhibit differences in emissions because of manufacturing and emissions-control design differences. Analyses of I/M data have shown that specific vehicle models have much lower or much higher emissions than average (Wenzel 1997). In general, vehicles with higher emissions exhibit much more test-to-test variability than lower-emitting vehicles (Bishop and Stedman 1996).

Many factors can contribute to variability in repeated emissions tests of the same vehicle. For example, failures of some emissions control system components (such as a partially degraded catalyst) can be intermittent and therefore result in higher emissions some of the time. Other sources of differences in the vehicle emissions test data underlying the MOBILE model arise from the testing process. For example, back-to-back emissions tests will vary because of differences in the measurement equipment, calibrations, and personnel (e.g., driving styles in tracking a target speed-time trace on a dynamometer).

Those and other factors create uncertainties in the statistical models fitted to the data that are the basis of MOBILE emissions factors. An example of the enormous scatter in the emissions test data underlying MOBILE is shown in Figure 4–2. The figure shows the test data used to estimate NO_x emissions for Tier 1 LDVs in MOBILE6. The solid line in the middle of the figure corresponds to 2 times the FTP standard; vehicles with emissions above this level are considered high emitters. Partly because of the lack of sufficient data, and partly because EPA assumes that high emitters are "broken" vehicles whose emissions are always high no matter how old the vehicles are, emissions for the high emitters are modeled as a constant (the upper dashed line). This emissions estimate is then adjusted with the Dayton IM240 data as described above. Below the solid line, the vehicles are considered to be normal emitters. For these vehicles, the basic emissions rate is modeled as a linear function of accumulated mileage (the lower dashed line). Clearly there is large uncertainty in the emissions data and consequently in the basic emissions rates for normal-and high-emitting vehicles estimated from these data.[2] These vehicle-to-vehicle differences are critical for some uses of MOBILE, but the scatter

[2]EPA has updated their original analysis of these data for Tier 1 NO_x emissions using additional data sets. The additional data reduces the confidence limits (although not the scatter) for the normal emission regression equation. However, there is still significant uncertainty in the estimated mean value for high emitters.

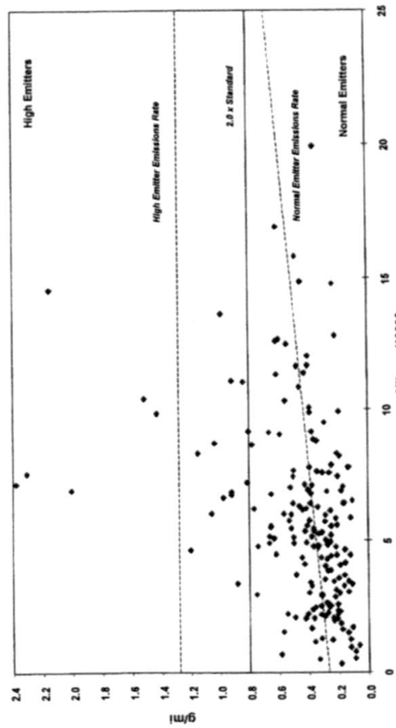

FIGURE 4–2 Vehicle emissions test data used to generate MOBILE6 NO_x basic emissions rates (g/mi) for TIER 1 light-duty vehicles. Source: EPA 1999p.

tends to get smoothed out in aggregated data, such as estimates of area-wide burdens. What does not get smoothed out is consistent bias, such as systematic underestimation of the emissions from high-emitters.

Incorrect Model Formulation

The emissions factors, emissions-factor adjustments, and estimates of emission control program effects in MOBILE6 and its predecessors are estimated from statistical analyses of available test data. The statistical models are chosen using both engineering considerations (to represent the physical process) and statistical considerations (such as selecting the model that produces the best statistical correlation). Uncertainties can arise from incorrect or inappropriate engineering and statistical models. Examples of incorrect physical-engineering models include the following:

- Neither MOBILE5 nor MOBILE6 includes road-grade effects, as these are not incorporated into the test data underlying the models. Road-grade influences vehicle emissions; higher emissions occur under vehicle load conditions such as steep road grades.
- Light-duty truck emissions are sometimes estimated from automobile test data, because of the lack of sufficient data for trucks. Although some light-duty trucks are used (and may emit like) passenger vehicles, other light-duty trucks are used regularly as working vehicles and frequently carry heavy loads; emissions from these working trucks are likely to be higher than automobiles of the same model and age.
- In MOBILE, the lack of sufficient data to estimate the effects of high emitters is sometimes filled in by assuming that high emitters behave as normal emitters do.
- In EPA's multistep adjustment of basic emissions rates for high emitters using Dayton IM240 data, one step is to estimate full IM240 emissions from fast-pass IM240 data (see Glossary for definition of fast-pass). This is done using second-by-second IM240 data from Wisconsin, so simulated fast-pass emissions can be compared to full IM240 emissions (EPA 1999g). The regression equation developed by EPA has the time (in seconds) of the fast-pass as one of the predictors of the full IM240 emissions; full IM240 emissions are assumed to be linearly related to the logarithm of fast-pass emissions. Given the variation in the speed-time trace of the IM240 data, there is no physical reason why fast-pass time should be linearly related to full IM240 (or to the log of full IM240) emissions. However, the coefficient for the fast-pass time is statistically significant in the regressions for all three pollutants (CO, VOCs, and NO_x). This appears to be because newer vehicles are much more likely to pass early in the test than older vehicles (Pollack et al. 1999a).

Inaccurate physical-engineering models can also arise from the failure to obtain correct data in a test program. Examples of that include the following:

- MOBILE5 speed-correction factors (SCFs) were determined from emissions of vehicles driving over a series of test driving cycles. The driving cycles were characterized only by average speed, although, as discussed in Chapter 3, there are many other factors in a driving pattern that also affect vehicle emissions. The SCFs used in MOBILE6 are facility-specific. For each facility, driving cycles were developed from analyses of real-world (instrumented vehicles) driving-pattern data. Although there is more aggressive driving in these cycles, and they are facility-specific, the cycles are still characterized for a given facility, by a single parameter—average speed. The development of modal (or second-by-second) emissions models (discussed in Chapter 5) will go a long way towards resolving these issues.
- The FTP cycle and the facility-specific speed correction factor cycles of MOBILE6 represent samples of a universe of driver and vehicle behavior. That universe may have low-frequency events with extremely high emissions. If these low-frequency high-emission events are not properly represented in the cycles used for MOBILE, an accurate picture of exhaust emissions will not be obtained.
- Some of the data critical to estimation of air-conditioning effects on emissions were not available for use in MOBILE6. These include the effect of vehicle speed, the time the compressor is on, the validity of the heat index as a measure of air-conditioner use, car occupant behavior in air-conditioner use, or the effect of the actual compressor torque.

Examples of incorrect statistical models include the following:

- In some analyses, the intercept of the statistical model is forced through zero, to match physical processes; this represents a trade-off between the correct statistical model and the correct physical model. Although such model alteration makes sense from an engineering point of view, it introduces bias into the resulting statistical model.

Two examples of this practice in MOBILE6 are the following:

—The determination of hot-running emissions from FTP test data; the test data are first transformed using logarithms, then the intercept is forced to be zero (EPA 1999k).

—MOBILE6 uses Dayton IM240 data to adjust FTP test data to account for the absence of high emitters in the FTP sample. The exhaust emissions rate adjustment is treated as an additive function of mileage, with zero increase in the adjustment at zero mileage (EPA 1999g).

- Logarithmic transformations are frequently applied to emissions data for fitting statistical models to bring the emissions data closer to the normal (Gausian) distribution, for which most statistical methods are developed. The transformed data may still be non-normal, and the translation of results back into the original scale may be hard to interpret.

In addition, the log transformation has the effect of reducing the influence of high values, which may be of greatest concern. Distributions of vehicle emissions typically show the majority of the emissions from normal emitters at relatively low levels, and a small fraction of the emissions at very high levels. These distributions with long right tails are much closer to lognormal than normal, although other distributions such as gamma have been fit as well (Zhang et al. 1994). Another reason that logarithmic models are commonly fit is that variability in vehicle emissions is higher at higher emissions levels; logarithmic transformations stabilize the variance. However, logarithmic transformations are usually inappropriate in analyses of data sets with both normal and high emitters. With logarithmic transformations, the effects of high emitters are minimized relative to normal emitters, whereas in the atmosphere the effects of high emitters are, in fact, much greater than normal emitters. In summary, logarithmic transformations offer some convenience in analysis and the opportunity to use simple statistical models, but at a cost of introducing potentially serious errors.

Figure 4–3 shows an example from MOBILE6 of an inappropriate use of a logarithmic model. The data in the figure show hydrocarbon (HC) emissions from model-year 1993 fuel-injected cars from a data set of Wisconsin IM240 second-by-second emissions; these data were used as part of the multistep process to adjust basic emissions rates using Dayton IM240 data (EPA 1999m). The Dayton data contain many fast-pass emissions tests, and the Wisconsin data were used to develop regression equations to predict fast-pass emissions (FHC on the horizontal axis) to full IM240 emissions (HC on the vertical axis). The graph on the left is a scatterplot of the full IM240 HC emissions against the fast-pass HC emissions; the top and right sides of the figure provide histograms of each of these variables individually. Clearly these emissions measurements are not normally distributed. The right graph shows the same data, but with a logarithmic transformation applied to both the full IM240 (LHCnz on the vertical axis) and fast-pass HC (LFHC on the horizontal axis) measurements. Both the scatterplot and the marginal histograms show that the logarithmic transformation is an over-transformation—it diminishes the effects of the high-

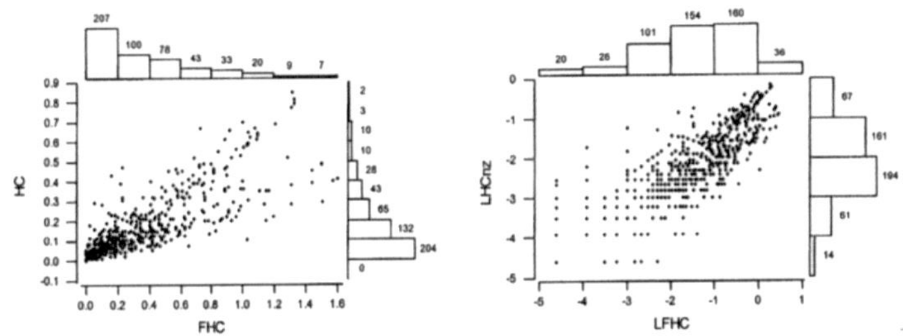

FIGURE 4–3 Example of inappropriate use of logarithmic transformation in MOBILE6 methodology for estimating deterioration rates. For the figure on the left, full IM240 hydrocarbon emissions are labeled HC and fast-pass hydrocarbon emissions are labeled FHC. For the figure on the right, logarithmic transformed full IM240 emissions are labeled LHCnz and logarithmic transformed fast-pass IM240 emissions are labeled LFHC. Source: Pollack et al. 1999a.

emitting vehicles and overemphasizes the effects of the lower-emitting vehicles.

Uncertainties in Vehicle-Activity Data

The MOBILE model estimates emissions factors in grams per mile (g/mi). Some of the uncertainties in these emissions factors have been described above. Local transportation and air-quality planners then obtain estimates of vehicle miles traveled (VMT). VMT estimates are multiplied by the MOBILE emissions factors to derive an estimate of total on-road vehicle emissions. Although VMT estimates are not part of the MOBILE model itself, they are essential to the use of the model, and mobile source emissions inventory results can be substantially affected by uncertainties in the VMT estimates. In addition, uncertainties in other inputs to MOBILE, such as average speed, vehicle age distributions, and vehicle mix, also contribute to uncertainties in MOBILE emissions factors and emissions inventories estimates with these factors. Reviewing the uncertainty in transportation inputs to the model is beyond the scope of this review, but work has been done in this area (e.g., Chatterjee et al. 1997).

PREVIOUS MOBILE SENSITIVITY AND UNCERTAINTY STUDIES

Complete characterization of MOBILE5 model sensitivity and uncertainties on emission-factors has not been conducted, although it is critical to accomplishing one of the specific tasks assigned to the committee. No comprehensive study has attempted to estimate the relative importance of a wide range of input and model uncertainties on output uncertainties. However, limited sensitivity and uncertainty analyses have been performed for both the current and earlier versions of MOBILE. These are described in this section. However, none of these estimates are truly comprehensive in their scope.

Sensitivity Analyses

EPA's Office of Transportation and Air Quality (OTAQ) and others have performed sensitivity analyses of recent versions of MOBILE. Although EPA's sensitivity analyses have been focused exclusively on model response to changes in input parameters only, other authors have reviewed model sensitivity to some parameters within the model. EPA first prepared a collection of graphs and tables showing sensitivity of MOBILE4 inputs to speed, temperature, and fuel Reid vapor pressure (RVP) (Shih

1990). For MOBILE5, EPA prepared a large collection of tables showing MOBILE5 emissions factors every 5 years from 1995 to 2020 by vehicle class and emissions mode as a function of input altitude, speed, operating-mode fractions, air-conditioning usage, and fraction of the fleet carrying extra load or towing a trailer (EPA 1998j).

In the first major outside review of MOBILE, Pollack et al. (1991) evaluated the sensitivity of MOBILE4 (and also EMFAC7E) to changes in both user inputs and model parameters and functions. Sensitivity of predicted emissions factors was evaluated in response to alternative fleet characterization (mileage accumulation, registration distributions, and diesel penetration rates), exhaust-emissions rates (emitter category proportions and deterioration rates), exhaust-correction factors (speed, temperature, and fuel volatility), I/M program parameters and tampering rates, and evaporative emissions-factor parameters.

Several studies have performed MOBILE5 sensitivity simulations. Heiken et al. (1994) assessed the sensitivity of model output to alternative formulations for exhaust-emissions rates, evaporative system pressure-purge test pass-fail rates, and evaporative basic emissions rates. Fox (1996), using Monte Carlo simulations, evaluated the sensitivity of MOBILE5 to uncertainties in seven key model inputs—minimum and maximum temperature, fuel RVP, average speed, fraction of VMT from catalyst and noncatalyst vehicles in cold-start operation, and fraction of VMT from catalyst vehicles in hot-start mode. Chatterjee et al. (1997) focused on key travel-related inputs to MOBILE5 (speed, VMT, vehicle classification, and operating-mode fractions) and assessed MOBILE5 sensitivity to these variables for conformity analyses.

Uncertainty Analyses

Several studies have been performed to assess uncertainties in specific MOBILE model components. Heiken et al. (1994) calculated confidence intervals for model components of particular interest to the American Petroleum Institute, the study's sponsor. These included IM240 to FTP correlations, running losses, and evaporative system pressure-purge test pass-fail rates. MOBILE5 SCFs have been the focus of two studies. Guensler (1993) and Guensler and Leonard (1997) assessed uncertainty in speed-correction factors (SCFs) using Monte Carlo analyses based on SCF regression errors. Chatterjee et al. (1997) performed statistical analyses of SCF regressions; confidence intervals for SCFs as a function of speed were then used to estimate uncertainty in emissions at typical roadway types and speeds (congested and uncongested freeways and congested arterials).

Kini and Frey (1997) derived estimates of uncertainties in basic emis sions rates and speed corrections combined for one specific MOBILE5 LDV technology group (port-fuel and throttle-body injection vehicles). Exhaust emissions errors for these vehicles only were estimated using a bootstrap technique applied to regression errors. Uncertainties for the emissions factors were found to be 20% to 40% for volatile organic compounds (VOCs), and 25% to 55% for NO_x; biases in the emissions factors were also found.

Uncertainties in all exhaust-emissions correction factors combined in the EMFAC7G model were determined by Pollack et al. (1999b). These correction factors (speed, fuel, off-cycle emissions, high emitter corrections, and variable start times), analogous to those in MOBILE, collectively introduce 20% to 40% uncertainty in the model's exhaust emissions rates. It is important to note, however, that in this and all uncertainty studies described here, the uncertainty estimates were derived considering only uncertainties in the statistical models fit to the available test data. Total MOBILE (and EMFAC) model uncertainty includes many other factors, and hence is likely to be substantially higher.

WHY UNCERTAINTY ANALYSES ARE NEEDED

A comprehensive uncertainty analysis for MOBILE could assess the relative contributions to model uncertainty of specific model parameters, data gaps, and model statistical and engineering formulations. Uncertainty attribution studies are used in other aspects of air-quality assessments, and have been effective in targeting areas for more detailed study. An uncertainty assessment of MOBILE is critical to provide model users with important information and to target future research.

At a minimum, the assessment should provide estimates of uncertainty for each of the model's output emissions factors, including contribution of each emissions factor to the overall uncertainty in the on-road inventory. For example, it might be the case that the emissions factors for motorcycle emissions have the greatest variability, but that their small contributions to the on-road inventory would not justify the cost of obtaining more test data.

An uncertainty profile is critically needed for guiding future mobile source modeling development. Specifically, an uncertainty profile should guide future emissions testing programs in generating data that would be most effective in improving the model.

Uncertainty analysis has other important uses. It provides information for interpreting model results and the appropriate use of those results, and can provide confidence to policy-makers making decisions based on associated analyses. Air-quality planners might wish to design their control programs to compensate for some margin of error, to have more confidence that air-quality goals for the future will be achieved. Further, an uncertainty analysis will provide information relevant to model accuracy, although it does not provide the actual accuracy of the model. As noted earlier in this chapter, the model can be very accurate even if the results are highly uncertain. Likewise, an uncertainty analysis might miss a key error, or the actual uncertainties in the model application may be greater than those tested ("uncertainties in the uncertainties"), leading to a larger error.

The problem, however, is that doing an assessment of uncertainty in the MOBILE model is difficult and resource-intensive. Although some methods have been developed and applied to estimate some of the components of uncertainty in MOBILE and related emissions models (described above under previous studies), there are intractable problems in estimating some of the components of the model's uncertainty. For example, it is difficult to assess the uncertainty arising from non-representative samples, or from incorrect model formulation. That a perfect assessment of uncertainty cannot be done, however, should not stop researchers from estimating the uncertainties that can be addressed quantitatively. To do so, EPA must archive all data used in the analyses leading to model formulation, and must carefully document all analysis approaches. As noted in this report, EPA has made substantial improvements in documenting modeling approaches, but is not archiving all data sets used in MOBILE6 development.

A critical unanswered question at the heart of the issues related to MOBILE's uncertainty and evaluation is how accurate the model needs to be to serve its various uses described in Chapter 2. This committee has not been able to identify any authoritative publication that says how close the model should be to reality. It is clear that EPA, working with the user community, should identify the level of accuracy desired for a particular application. For example, if it were determined that the current version of MOBILE is accurate enough to fulfill all of its roles (which the committee deems it is not), little further work in that direction is necessary. On the other hand, if a significant reduction in model uncertainty is warranted, then considerable work is suggested, and possibly a new design if the current approach is too limited (as is concluded here). There will not be a single answer, as various applications demand different levels of accuracy. Designing future emissions models should take into account the accuracy required for different applications.

INTRODUCTION TO EVALUATION

Considering the important applications of MOBILE, it is critical that MOBILE reflects actual emissions of the fleet of on-road vehicles. Much of the knowledge and information used in creating MOBILE, have come from FTP-based testing of small numbers of vehicles. This information is weighted to reflect the amount of cold and hot starts and stabilized running conditions estimated to be occurring in the real world, as described in Chapter 3. This approach may not produce estimates representative of the in-use fleet because of such factors as vehicle-recruitment bias (typically well over 75% of the selected vehicle owners refuse to participate in FTP testing), inaccurate and outdated characterization of the average real-world driving cycle, and small databases. For instance, the FTP driving cycle is based on driving conditions in Los Angeles in the 1960s.

EPA has done little to test the accuracy of MOBILE other than conduct a recent comparative analysis and adjustment to state I/M test data, which is subject to inaccuracy in reflecting average real-world fleet emissions (EPA 1999g; EPA 1999m). Many different independent techniques could be used to help evaluate the accuracy of MOBILE. A wide range of studies, using a variety of techniques, can already provide more direct insight into the accuracy of MOBILE. Some of the results of these studies are presented in the following sections. In general, these studies indicate that MOBILE substantially underestimates VOC emissions, while NO_x emissions appear to be more accurately predicted. This results in a discrepancy in the VOC to NO_x ratio when comparing ambient observations to model predictions.

MOBILE'S SENSITIVITY TO VARIATION IN DRIVING AND STARTS

The New York State Department of Environmental Conservation (NYDEC) in 1994 and 1995 developed mobile-source inventories using microscale driving-cycle data (speeds, ambient temperature, engine operating mode, and VMT by hour) for a wide range of counties in the state (Keenan and Escarpeta 1994). This study can be used to help assess the sensitivity of MOBILE's vehicle emissions rates to changes in the driving cycle and ambient temperature. The results in Figure 4–4 illustrate MOBILE's VOC and NO_x emissions ratio sensitivity to changes in traffic conditions. On the left, measured average speeds are shown for 17 different counties. On the right, MOBILE-predicted ratio of VOC to NO_x corresponding to the local driving-cycle data on two different days (at two different temperatures) are shown. VOC to NO_x ratios are indicated to rapidly rise as the VMT-weighted average vehicle speeds fall for counties ranging from very rural to highly urbanized. The figure shows that model estimated emissions are very sensitive to details of local driving conditions.

The NYDEC study also found that MOBILE's driving-cycle default value for the engine thermal state operating mode (cold- and hot-start portion of VMT) was not representative of any of the wide range of traffic con ditions studied (very rural to highly urbanized counties in New York State), as indicated in Table 4–1. Because cold starts account for a significant portion of the VOC and carbon monoxide (CO) emissions of the total driving cycle, underestimating the frequency of cold starts can significantly contribute to the underprediction of VOC and CO emissions by MOBILE.

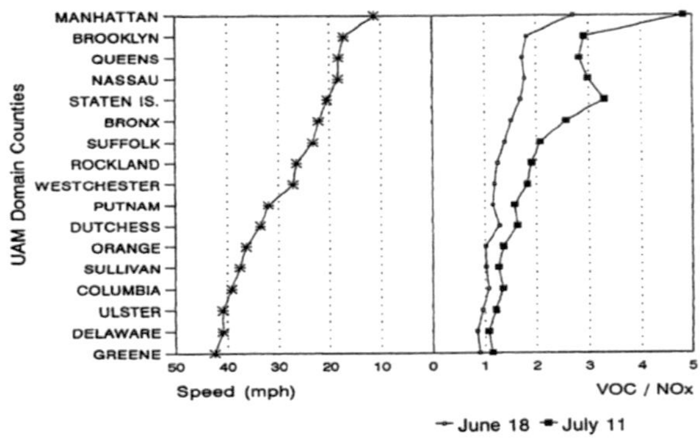

FIGURE 4–4 Emissions ratios derived from NYDEC study. The left panel shows the effects of increased population density on average driving speed for 17 counties. The counties start with the lowest population density at the bottom (Greene) and end with the highest at the top (Manhattan). The right panel shows the effect of applying the average speed data to MOBILE to predict the VOC/NO_x mass ratio. These emissions show a large sensitivity to the details of local urban driving. Source: Keenan and Escarpeta 1994.

In summary, the NYDEC study illustrates that slight inaccuracies in the assumptions, default values, or specific input of driving-cycle parameters to MOBILE (particularly speed input and the hot- and cold-start mode split) can lead to significant underprediction of real-world fleet VOC emissions. This indicates the need for input parameters to be based on observations of local conditions. While the NYDEC study was based on MOBILE5, the conclusions continue to be applicable to MOBILE6, although MOBILE6 has been improved by separating cold start emissions from running emissions. A fundamental change to modal (instantaneous) emissions modeling would be necessary to ensure that effects of local driving conditions are accurately reflected on emissions.

TABLE 4–1 The Percent of Total VMT Allocated to Cold Starts and Hot Starts for Real-World Measurements Compared With MOBILE Predictions

	VMT Weighted Daily Average	
Area	Cold Starts	Hot Starts
New York State (average)	25.6%	10.4%
Upstate New York (rural areas)	27.1%	12.2%
Downstate New York (metropolitan areas)	22.8%	9.5%
MOBILE Default	20.6%	27.3%

Source: Keenan and Escarpeta 1994.

STATE EMISSIONS INSPECTION TESTING

State I/M testing programs provide vehicle emissions test data for a large population of the real-world fleet. EPA has recently used data from selected state IM240[3] tests to correct deterioration rates for running emissions in the upcoming MOBILE6 model (EPA 1999m). EPA recognizes some limitations of IM240 data, such as the absence of cold starts and shortcomings in the driving cycle. But the agency argues that this technique represents a census of all vehicles and, thus, is relatively unbiased. Figure 4–5 shows EPA's formula fitted adjustment to FTP-based testing results using the IM240 data. The revised model, based on IM240 data, results in a large change in predicted emissions of about +150% for VOCs, +50% for CO, and +30% for NO_x for 1988–1993 port-fuel injection cars, as illustrated in this figure.

EPA indicates that existing IM240 data still suffer from lack of uniform preconditioning. If every vehicle were preconditioned with a full IM240 cycle prior to the IM240 test cycle, the results could be useful in evaluating the predicted MOBILE6 emissions rates. However, there are other serious limitations to state IM240 data that bias it away from true real-world

[3]The IM240 test differs from an idle or tailpipe test in that the emissions are measured while the vehicle is driving on a dynamometer. The test driving cycle that includes accelerations, cruise, idle and deceleration, operated is intended to assess emissions under conditions more similar to actual driving conditions that a test of an idle vehicle.

emissions of the overall fleet. Thus, even a preconditioned set of I/M data is not ideal for testing MOBILE. Among limitations of the IM240 test as a measure of overall fleet emissions are that the test does not

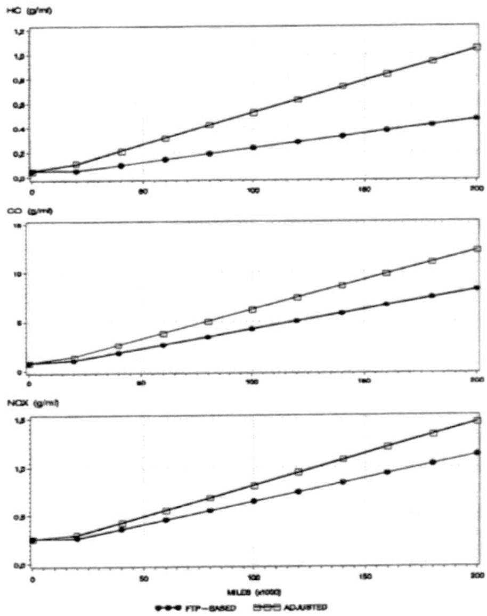

FIGURE 4–5 EPA adjustment to MOBILE6 FTP data based on the analysis of IM240 data. Source: EPA 1999g.

- account for new vehicles because they are exempt from IM testing;
- account for unregistered vehicles;
- reflect the level of emissions between test cycles;
- reflect tampering after the test;
- reflect emissions from vehicles operating in the "prior to preconditioning" state;
- reflect vehicles that "disappear" from the I/M program after failing the test but continue to operate in the urban area;
- reflect vehicles that register outside the I/M boundary to escape testing requirements but continue to be operated in the urban area; and
- account for cold-start effects.

Additionally a fully preconditioned IM240 data set would not provide any measure of cold-start emissions. Two elements are necessary for the accuracy of the MOBILE model regarding cold starts. The first is a reliable measurement of the magnitude of cold-start emissions, including the dependence on mileage and age for both normal and high emitters, and the effect of soak time. The second is an accurate measure of the population of vehicles with high-start emissions. In MOBILE6, EPA has assumed that the fraction of high emitters for start emissions is the same as the fraction of high emitters for running emissions. High emitters also might have catalyst malfunctions, including increases in light-off temperature. An increase in light-off temperature would have more of an effect on start emissions than on running emissions. The IM240 adjustment that EPA made for MOBILE6 was one step toward improving estimates of deterioration rates and providing a more accurate reflection of real-world conditions. However, this improved estimate directly applies only to running emissions. EPA should consider methods for improving its estimates of real-world start emissions in future versions of MOBILE.

REMOTE SENSING

Remote-sensing measurement of emissions of vehicles while they are in use provides another approach to generating a large database of real-world emissions that can be used to test the accuracy of MOBILE. Remote sensing has the advantage of being able to capture a snapshot of a very large real-world fleet without the disadvantages of using IM240 data, which have been previously mentioned. Its main disadvantages are that it is does not provide an accurate measurement at low emissions levels and it reflects vehicles typically in a single mode of operation.

A recent comprehensive analysis of Arizona IM240 data using a very large remote-sensing database confirms some of the limitations of using IM240 data to reflect real-world fleet emissions reductions from I/M pro grams (Wenzel 1999). The study included 4 million readings of 1.2 million vehicles. The study compared the MOBILE and IM240s estimate of the emission reductions resulting from an I/M program versus estimates made from the remote-sensing data for vehicles subject to IM240 and remote sensing tests. The results, shown in Table 4–2, indicate emissions reductions from I/M programs calculated from IM240 data are less than the reductions predicted by MOBIILE. Using remote sensing to estimate emissions reductions from I/M produces the smallest estimated emissions reductions. The discrepancies, particularly between results from IM240 and remote sensing data are likely due to

- about 33% of vehicles failing I/M testing (high emitters) do not return for retest (disappear), and 60% of those that fail continue to operate in the airshed after 6 months; and
- repaired vehicles have a high deterioration rate.

ROADSIDE INSPECTION

Roadside pullover studies in which a random sample of vehicles are pulled off the road and subject to a loaded-mode emissions test[4] could perhaps offers the best direct measure of the overall real-world fleet emissions rates if a sufficient sample size is collected. Because on-road vehicles are selected at random and emissions are measured under actual real-world conditions, it does not have most of the disadvantages of IM240 testing and remote sensing previously mentioned.

TABLE 4–2 Comparison of Emission Reduction Estimates for an I/M Program Based on MOBILE, IM240, and Remote Sensing Data

Emissions Reduction Method	CO	VOC	NO_x
MOBILE5 Prediction	16.2%	16.9%	16.7%
Arizona IM240 Data Analysis	14.5%	14.0%	7.1%
Arizona Remote-Sensing Data Analysis	7%	11%	Not measured

Source: Wenzel 1999.

[4]A loaded-mode test is one that puts a car through a simulated driving cycle on a dynamometer.

The California Bureau of Automotive Repair (BAR) conducted random road-side pullover inspections of over 27,000 vehicles in 1997–1999. The objectives of the study are to help characterize fleet emissions and establish a baseline for evaluating California's I/M program effectiveness. BAR recognized the potential inaccuracies of directly using I/M program data to assess real-world effectiveness of I/M. BAR is also conducting a similar series of tests after implementation of an enhanced I/M program in California to accurately assess the effectiveness of this program. When the results and analysis of this test program become available, they might provide the best insight into the accuracy of emissions models such as MOBILE.

AMBIENT AIR-QUALITY MONITORING AND MODELING

Using MOBILE-generated emissions data in airshed modeling and comparing the results with measured air-quality data offers yet another approach to testing the accuracy of MOBILE emissions predictions. An extensive ozone-modeling and emissions-inventory study comparison was made for the South Coast Air Basin in the Los Angeles area in 1987 (Chico et al 1993; Harley et al 1993b; Wagner and Wheeler 1993). This study used California's EMFAC mobile-source emissions-inventory model. EMFAC has produced lower estimated emissions than previous versions of MOBILE, but the emerging MOBILE6 is expected to compare relatively closely with previous versions of EMFAC. Thus, the California study has relevance to evaluating the accuracy of MOBILE. This study found that airshed modeling substantially underpredicted ozone levels. Additionally, it was determined that when the on-road mobile-source VOC emissions, predicted by EMFAC, were multiplied by a factor of 2.5 and the airshed model rerun, the airshed model predictions of ozone closely matched ambient measurements. Figure 4–6 shows the comparison between ambient observations and the airshed model run for both levels of VOC emissions.

The South Coast study also compared CO to NO_x ratios and VOC to NO_x ratios derived from the EMFAC-airshed model predictions with ambient measurements in the Sherman Way tunnel in Van Nuys. Table 4–3 shows that the two measured values are similar, further indicating that the emissions model underpredicts CO and VOC emissions from motor vehicles by about a factor of 2.

A more recent Desert Research Institute (DRI) study of ambient air-quality measurements versus emissions model estimates for Los Angeles (Zielinska et al 1999) indicates a deviation in the VOC to NO_x ratios similar to the findings in the 1987 South Coast study. The results of this study found that the EMFAC model underestimates the VOC to NO_x ratio of emissions by a factor of 2.15 compared with ambient data collected in Los Angeles in 1995. MOBILE6 will possibly make these ambient and emissions inventory discrepancies even worse. This conclusion is based on emissions data in EPA's paper (1998k) on possible regulatory action with respect to Tier 2 vehicle emissions and gasoline sulfur standards, in which EPA made adjustments to MOBILE5b to simulate the expected MOBILE6. MOBILE6 was unavailable to the committee to confirm if this conclusion is correct.

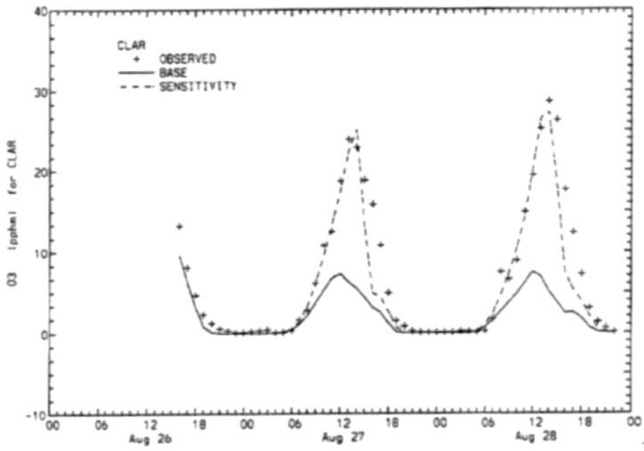

FIGURE 4–6 Comparison of airshed model predictions of diurnal ozone concentrations with observations of ambient ozone concentrations for two different VOC emissions levels. Model-predicted VOC emissions are multiplied by a factor of 2.5 in the sensitivity case to improve the fit to the ambient ozone concentration profile. Observations are at Claremont (CLAR), California. Source: Fujita 1999.

The South Coast and DRI studies are thus consistent with other independent techniques (IM240 and remote sensing) in demonstrating that EMFAC (and, likely MOBILE as well) underestimates VOC emissions and, in particular, that it results in a significantly inaccurate estimate of the VOC to NO_x emissions ratio of the real-world fleet.

TABLE 4–3 Ratio of Measured to Modeled Emissions, Using EMFAC7E Model Results, for Overall Ambient Measurements and Tunnel Studies

Ratio	Ambient Data	Sherman Way Tunnel
$\dfrac{\left[\dfrac{CO}{NOx}\right]_{Measured}}{\left[\dfrac{CO}{NOx}\right]_{Modeled}}$	1.5	2.1
$\dfrac{\left[\dfrac{VOC}{NOx}\right]_{Measured}}{\left[\dfrac{VOC}{NOx}\right]_{Modeled}}$	2.5	2.2

Source: Fujita 1999.

TUNNEL STUDIES

The analysis of air samples in highway tunnels has been used for several years as a means of measuring vehicle emissions and testing model predictions (e.g., Pierson et al. 1990). Tunnel studies capture the emissions from a large number (typically thousands) of in-use vehicles, thus providing measurements of the fleet's average emissions. Generally the vehicles are operating in a hot-stabilized cruise mode with average speeds from around 25 to 70 miles per hour (mph). Some tunnels have a significant grade.

A report published by the Coordinating Research Council (CRC) (Gertler et al. 1997) summarizes results from a 1995 study in five different tunnels in Boston, New York City, Phoenix, and Los Angeles as well as from several previous urban tunnel studies. The CRC tunnel study results are summarized in Table 4–4, where the average emissions factors for CO, VOCs, and NO_x are given, as well as the ratios of several emissions factors. The sampled fleets were largely light-duty gasoline fueled vehicles. The data for the Fort McHenry and Tuscarora tunnels were obtained for a mixture of light-duty and heavy-duty vehicles and were analyzed to extract the light-duty component reported in the table.

The third column of the table shows a range of about 2-fold in the average speed of the vehicles in the different tunnels. Emissions of CO, VOCs, and NO_x differed substantially among the tunnels, typically about a factor

TABLE 4-4 Summary of Results of Nine Tunnel Studies of On-Road Vehicle Emissions

Site	Year	Avg. Vehicle Speed (mph)	LDVs (%)	CO (g/mi)	VOCs (g/mi)	NO_x (g/mi)	CO/NO_x (g/g)	VOC/ NO_x (g/g)	CO/CO_2 (g/g)
Van Nuys	1995	43.6 ± 5.4	97.2	13.9	1.04	1.25	11.2	0.83	0.043
Sepulveda	1995	45.7 ± 8.0	97.2	9.6	0.89	1.15	8.35	0.78	0.026
Deck P-S	1995	59.7	94.5	6.4	0.87	1.20	5.32	0.83	0.020
Lincoln	1995	27.1 ± 4.3	87	7.3	0.98	2.03	3.60	0.49	0.017
Callahan	1995	26.4 ± 5.2	96.3	5.4	0.63	1.11	4.87	0.57	0.020
Van Nuys	1987			21.0	2.7[a]	1.59	13.6	1.70[a]	N/A
Caldecott	1994			N/A	N/A	N/A	10.3	0.55[a]	N/A
Fort McHenry[b]	1992		100	6.4	0.62	0.81	7.80	0.76	0.023
Tuscarora[b]	1992		100	4.9	0.29	0.39	12.0	0.74	0.020

[a] Results are for total HC emissions.

[b] The LDV emissions were separated from HDV emissions by data analysis. These tunnels are characterized as high-speed interstate highway tunnels.

Source: Gertler et al. 1997.

of 4. These differences can be accounted for by the differences in the ages and modes of operation of the vehicles being tested. However, there is much less variation in the ratios of some of the pollutants, notably the VOC to NO_x and CO to CO_2 ratios. The CRC report discusses some inconsistencies between these data and both the MOBILE and California's EMFAC emissions models.

Although tunnel data provide only a snapshot of vehicle emissions, they can be valuable and should continue to be used in testing the accuracy of MOBILE and examining the effects of fuels, operating mode, and fleet composition. However, vehicle operation in tunnels tends to significantly deviate from average real-world conditions. Tunnel traffic tends to have higher speeds, less stop-and-go driving, and less loaded-mode operation than average real-world urban conditions.

CHEMICAL-MASS BALANCE

Chemical-mass balance (CMB) is another approach for evaluating emissions model estimates with ambient observations. CMB uses a sophisticated set of chemical fingerprints derived from speciation of source emissions, which are apportioned mathematically to ambient air samples. An oxidant assessment study in southeast Texas (Fujita et al. 1995) included extensive comparison CMB estimates versus MOBILE-estimated VOC and NO_x emissions.

This study distinguished contributions from liquid gasoline and gasoline vapor from vehicle exhaust in the ambient atmospheric measurements. The findings of this study and another CMB study (Korc et al. 1995) counter arguments that it is evaporative emissions that account for MOBILE's underprediction of VOCs. The major conclusions of these studies are

- The sum of ambient liquid gasoline, gasoline vapors, industrial and compressed natural gas contributions agrees reasonably well with and validates the corresponding MOBILE emissions inventory estimates.
- The discrepancies between CMB ambient- and emissions-derived VOC to NO_x ratios and ambient- and emissions-derived acetylene (a major tracer of fingerprint of motor vehicle exhaust) at the Clinton site suggest that the absolute amount of on-road mobile-source exhaust VOC emissions were substantially underestimated by MOBILE.
- The average ratio of CMB-derived ambient VOC emissions from mobile sources compared with those estimated from MOBILE at the Clinton site was 2.3.

The latest state-of-the-art CMB study was conducted in the Denver area in 1996 and 1997 under the Northern Front Range Air Quality Study (NFRAQS). The study was directed toward evaluating participate matter (PM) emissions and sources; it indicated that PART5 greatly underestimates the contribution of gasoline vehicles compared with diesel vehicles. Tables 3–10 and 3–11 show this underestimation of emissions rates for LDVs by PART5. The study also found a very large contribution of start emissions and high emitters to the total emissions from gasoline vehicles. Table 4–5 provides a summary of pertinent NFRAQS results. Underestimating emissions from starts and high emitters may be a significant cause of MOBILE's underprediction of other emissions (CO and VOCs) found in various studies.

Watson et al. (in press) contains a recent summary of CMB studies. These studies tend to show that the relative contributions of mobile source VOC emissions to the total inventory determined by CMB are two to three times higher than those estimated using mobile source emissions factor models such as MOBILE and EMFAC.

FUEL-BASED APPROACH TO EMISSIONS ANALYSIS

Another approach to evaluating mobile-source emissions estimates from MOBILE is to compare MOBILE's results with those estimated through a fuel-based approach. The remote-sensing and tunnel-study methods described above measure exhaust emissions under operating conditions, but, unlike a dynamometer test, they do not measure the emissions on a gram per mile basis. The remote-sensing and tunnel studies measure the emissions concentrations of VOCs, CO, and NO_x relative to the concentration of the combustion product carbon dioxide (CO_2). Carbon dioxide is the major carbon-containing product of fuel combustion, and thus the CO_2 emission provides a measure of the amount of fuel burned. For improved accuracy, a correction is applied for the carbon in the VOCs and CO emissions. The determination of the concentrations of VOCs, CO, and NO_x relative to CO_2 provide measurements of these emissions relative to the amount of fuel consumed. These fuel-based emissions factors, measured in grams per gallon (g/gal), vary much less with changes in the vehicle mode of operation than do the travel-based (g/mi) emissions factors (Singer and Harley 1996; Singer et al. 1999). This might give the fuel-based emissions method an advantage over the travel-based method because the latter might not accurately represent the variations in the driving cycles of urban areas. Because the fuel-based emissions approach is less sensitive to the details of the vehicles' operation (speed and acceleration), this method is less susceptible to inaccuracies derived from the MOBILE model's failure to represent realistic urban vehicle operation.

TABLE 4–5 Comparison of CMB and PART5 Particulate Emissions from NFRAQS Study

Vehicle Category	CMB	PART5
Cold Starts	32.5%	3/4
Non-Smokers	7.5%	29.0%
High PM Emitter	31.3%	3/4
Diesel	28.8%	71.0%

Source: Fujita et al. 1998; Watson et al. 1998.

The fuel-based emissions inventory is a concept developed from measurements of vehicle emissions on a gram per gallon of fuel. If the emissions amounts in terms of g/gal for a representative on-road fleet are known, then the fleet-wide emissions rate using the fuel consumption rate in gallons per unit of time can be calculated. Fuel consumption is determined from fuel sales records. Knowledge of the fleet composition and the fuel economy of the different types of vehicles is also required to estimate the emissions inventory.

As described in previous sections, the MOBILE model employs a travel-based method to develop emissions inventories. This requires knowledge of the emissions levels (g/mi) for different modes of driving or for representative driving trips, vehicle activity or use, and the fleet composition. Both the travel-based and fuel-based methods require information on the in-use fleet composition, ambient temperature, and other factors that affect emissions and vary with geographical region.

Singer and Harley (1996) developed a fuel-based CO inventory for the South Coast Air Basin in California and compared the results to California's MVEI7F model. The fuel-based CO inventory estimate was of a factor 2.2 times larger for cars and 2.6 times larger for trucks than the travel-based model. In a second study used more than 60,000 remote-sensing measurements made at 38 sites in Los Angeles between May and October 1997, to develop CO and VOC inventories using the fuel-based method (Singer and Harley 2000). Their estimates for the on-road, fleet-stabilized exhaust emissions from cars and light- and medium-duty trucks are larger than the California MVEI7G model by factors of 2.4±0.2 for CO and 3.5± 0.6 for VOCs. Similar tests of the MOBILE model emissions inventories should be made to test the model's consistency and accuracy. The fuel-based approach to emissions inventories is a promising method that can be used to reduce the uncertainties in emissions predictions.

SUMMARY OF FINDINGS AND RECOMMENDATIONS

In this chapter, quantitative estimates of MOBILE's overall accuracy and uncertainty were not provided. The data are not available to do so with any confidence. Some studies have identified significant discrepancies in MOBILE predictions, but a complete assessment is not possible. Specific findings and recommendations follow.

Findings

1. EPA has done very little to use existing independent techniques to test the overall ability of MOBILE to accurately estimate real-world emissions of the overall on-road fleet. Others have applied many different techniques that test the accuracy of MOBILE. Most studies have found that MOBILE5 and earlier versions are substantially underpredicting VOC emissions (approximately by a factor of 2) and, to some extent, underpredicting CO and the PM contributions of gasoline-powered vehicles to the overall emissions. This is in contrast to NO_x emissions, which appear to be more accurately estimated.
2. At present there is an inadequate understanding and quantification of the sources of uncertainties in MOBILE. These uncertainties arise from small and nonrepresentative emissions data, statistical analyses of these data, and assumptions that underlie and define the MOBILE model's algorithms and predictions. Quantification of uncertainties is critical for understanding the weaknesses in the model, and identifying the most critical needs for further emissions test data.
3. A critical unanswered question at the heart of issues related to MOBILE's uncertainty and evaluation is how accurate the model needs to be to serve its various uses described in Chapter 2. It appears that EPA has not determined a desirable level of accuracy for MOBILE. It is clear that EPA, working with the user community, should determine the level of accuracy needed, and plan accordingly. For example, if it was determined that the current version of MOBILE is accurate enough to fulfill all of its roles (which it is not), little further work in that direction is necessary. On the other hand, if a significant improvement in model accuracy is demanded, then considerable work is suggested, and possibly a new design if the current approach is too limited (as is concluded here). There will not be a single answer, as various applications demand different levels of accuracy. Designing future emissions models should take required accuracy for different applications into account.
4. EPA has provided only limited sensitivity analyses for earlier versions of MOBILE. Although the primary source of uncertainty in MOBILE

output is in the underlying emissions data and data analyses used to generate the model formulations, uncertainty in model *inputs* can also create uncertainty in model outputs. It is important that air-quality planners understand the sensitivity of model outputs to uncertainty in model inputs such as average speed, ambient temperature, fuel volatility, and I/M program parameters.

5. A dominant cause of MOBILE's underprediction of real-world emissions appears to be related to the driving-cycle testing protocol (because the FTP-based driving cycle is not representative of current driving patterns) and associated adjustment factors and default values forming the basis for MOBILE. In the real world, slower speeds, heavier-loaded operating conditions associated with more congested stop-and-go driving conditions, and a more dominant role of cold starts, all of which produce enriched engine-operating conditions, appear to be the key factors in explaining the discrepancy of the real-world driving cycle compared with the average driving cycle reflected by MOBILE. This observation indicates the essential need to move toward a true modal modeling approach.
6. Other important factors that appear to be significant sources of error in MOBILE are recruitment bias in vehicles that are FTP-tested and small databases for sensitive parameters.
7. Although corrections have been made to many parts of the emerging MOBILE6 model to improve its accuracy, it appears the end result might deviate even more than past versions of the model with respect to the VOC to NO_x ratios. This will create, among other things, greater uncertainty in ozone modeling.

Recommendations

1. EPA should assess the levels of accuracy needed to fulfill its regulatory responsibilities and required for specific applications of the MOBILE model. EPA should compare the needed accuracy to the accuracy of the MOBILE model, and identify specific elements of the model that contribute most to its inaccuracy. EPA should use the results of such an assessment to help guide the development of the next generation of models that would have improved accuracy in critical model components.
2. Enhanced model evaluation studies should begin immediately and continue throughout the long-term evolution and development of mobile-source emissions models. These studies need to be conducted to reduce gaps between model-predicted emissions and the resulting air quality, and also to reduce gaps between model-predicted emissions reductions from control programs, such as vehicle I/M programs, and those that actually occur in implementation. The evaluation should include (but not be lim

ited to) field observations, including tunnel studies; remote-sensing measurements; source-receptor modeling; roadside pullovers; and air-quality monitoring and modeling; vehicle emissions testing data from vehicle I/M programs; and other vehicle emissions tests. Evaluation studies should be done with oversight and guidance from an independent body, including technical experts, and should be undertaken in tandem with the sensitivity and uncertainty studies suggested in the next recommendations.

3. Rigorous sensitivity analyses should be performed for all model inputs and provided as part of user guides for MOBILE6 and all future versions of MOBILE. From these sensitivity analyses, EPA should provide guidance to transportation and air-quality planners on the most critical model inputs affecting model results.
4. EPA, along with other agencies and industries, should undertake the necessary measures to conduct quantitative uncertainty analyses of the mobile-source emissions models in the modeling toolkit (discussed in Chapter 6), especially the MOBILE model. Future versions of the MOBILE model and other models in the toolkit should be developed to facilitate uncertainty analyses. Results of the uncertainty analyses should be used to guide research plans for obtaining additional test data that would increase the accuracy of the model.

5

Alternative Mobile-Source Emissions Modeling Techniques

N ADDITION TO MOBILE, numerous other computer models that estimate vehicle emissions have been developed over the years. Several of these models are regional in nature, simulating emissions over large areas, and others are far more microscale, simulating emissions along a corridor, at an intersection, or from an individual transportation project. This chapter briefly describes these alternative mobile-source emissions models, with a focus on defining key differences between MOBILE and the alternatives. Further, this chapter builds upon the transportation and emissions-model integration issues introduced in Chapter 2, defining how mobile-source emissions inventories can be generated at different levels of detail.

CALIFORNIA AIR RESOURCES BOARD MOTOR-VEHICLE EMISSIONS INVENTORY SUITE

The Clean Air Act allows California to adopt more restrictive automobile and fuel standards. This motivated the California Air Resources Board (CARB) to develop a mobile-source emissions model that more closely reflected their standards. The California model also integrates data sets for region-specific fleet characteristics and travel, allowing it to contain the activity data necessary to estimate both emissions factors as well as inventories. Below is a brief description of their motor-vehicle emis sions inventory (MVEI) suite, as well as a discussion of the differences between the California approach and MOBILE.

Overview and Recent History of MVEI

The current version of the CARB mobile-source emissions inventory model is designated as MVEI7G (CARB 1996b). This is actually a suite of models consisting of the following components, their relationship is shown in Figure 5–1:

- ***CALIMFAC*** is used to compute the basic emissions rates for light-and medium-duty gasoline-powered vehicles. The output of CALIMFAC is a set of regression equations giving the emissions rates as a function of calendar year for these vehicles.
- ***WEIGHT*** calculates the distribution of vehicles, starts, and vehicle miles traveled (VMT) by model year and vehicle category.
- ***EMFAC*** calculates all emissions rates (exhaust, evaporative, and tire and brake wear) for a specified calendar year for all vehicle types. These rates are computed as a function of vehicle speed and temperature.
- ***BURDEN*** calculates the emissions inventory (tons/day) for a specified county, air basin, or the entire state.

One significant difference between the structure of the MOBILE program and the CARB model is the addition of an emissions inventory module, BURDEN. The U.S. Environmental Protection Agency's (EPA's) MOBILE model is designed to compute the emissions rates from vehicles. This is the function performed by CARB's EMFAC module. MOBILE has default distributions for the fraction of VMTs by various vehicle classes and model years.

The new version of the CARB emissions factor model is EMFAC2000. This model has a large amount of area-specific data to compute vehicle activity in various areas of the state. Preliminary results from this model show significant increases in emissions, with statewide inventories for on-road vehicles increasing 68% for CO, 78% for VOCs, and 93% for NO_x.

Emissions Categories

MVEI provides estimates of gaseous and particulate emissions. Gaseous species are volatile organic compounds (VOCs), carbon monoxide (CO), oxides of nitrogen (NO_x), and sulfur dioxide. VOCs can be expressed as total hydrocarbons (HC), nonmethane hydrocarbons, or reactive organic

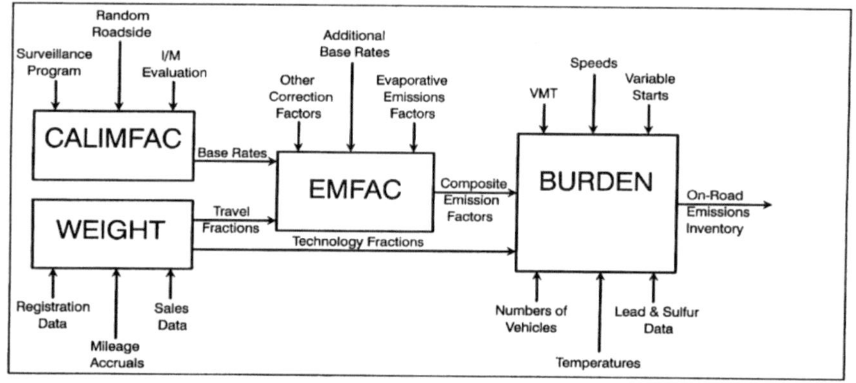

FIGURE 5-1 Structure of California Air Resources Board MVEI Model. Source: GARB 1996a.

gases. Only total hydrocarbons are actually computed; adjustment factors are used to obtain the other VOCs measures. (The differences among the various terms used for classifying organic compounds is shown in Appendix B.) MVEI computes emissions of lead, total exhaust particulates, and particulate emissions from tire wear and brake dust. It also computes fuel consumption. It does not compute refueling emissions. In California, refueling emissions are considered a stationary source and are handled as a separate part of the inventory. EMFAC2000 will include computations of carbon dioxide emissions.

In EMFAC2000, emissions are computed for passenger cars, eight weight classes of trucks, school buses, urban buses, motorcycles, and motor homes. Vehicle classes are subdivided into gasoline-fueled, diesel-fueled, and electric. Gasoline-fueled vehicles are further subdivided into catalyst and noncatalyst.

Basic Operation of CARB Models

The overall approach for EMFAC2000, MOBILE5, and MOBILE6 is very similar. All these models rely upon regression analyses of data sets to get basic emissions rates and correction factors. There is some sharing of data between the two models, but generally the main data sets—for exhaust emissions of light-duty vehicles (LDVs)—have been kept separate because of the differences in emissions standards between California vehicles and those sold in the rest of the country.

Because the approaches of the California model are similar to MOBILE, the potential accuracy limitations are the same. The Coordinating Research Council has sponsored a detailed study of the accuracy of the EMFAC module in MVEI7G (Pollack et al. 1999b). That review has noted several data and analysis limitations, which are similar to the criticisms leveled against the MOBILE models. Those include the need for better data on high emitters, heavy-duty vehicles (HDVs), start emissions, particulate emissions, and evaporative emissions. It also noted the need to improve estimates of the effects of air-conditioning, inspection and maintenance (I/M), and on-board diagnostic (OBD) on emissions.

Key Technical Differences Between EMFAC2000 and MOBILE6

Starting with EMFAC2000, the CARB model for light-duty vehicles (LDVs) will be based on a new cycle called the "unified" or LA92 cycle (Carlock 1999). The data for this cycle have been obtained on recent sur veys, which have measured emissions on both the unified cycle and the Federal Test Procedure (FTP) cycle. Users of EMFAC2000 will have the option of obtaining emissions based on the FTP or the unified cycle. The unified cycle emissions should be closer to real-world driving results. MOBILE6 will continue to report FTP emissions, adjusted by IM240 data.

The new CARB model will continue to use five emissions categories (normal, moderate, high, very high, and super) for light- and medium-duty vehicles. This contrasts with the decision to reduce the emissions categories from four in MOBILE5 to two in MOBILE6. The emissions category boundaries for MOBILE6 and EMFAC2000 are compared in Table 5–1. The category definitions in EMFAC2000 provide more distinction among emissions. However, these definitions increase the data requirements for getting a reasonable sample in each category.

Much of the development of EMFAC2000 has been devoted to the inclusion of area-specific activity data for various regions of California. Area-specific vehicle registration data, mileage accumulation data, vehicle age distributions, and data on VTM are provided for different areas of the state. Area-specific data on temperature and humidity profiles are also provided.

EMFAC2000 will use trip-based speed-correction factors (SCFs) rather than facility-specific speed correction factors that are planned for MOBILE6. The trip-based SCFs are appropriate for an emissions inventory, but link-based SCFs, which should be given by the facility-specific cycles in MOBILE6, are more appropriate for applications such as conformity determinations (Neimeier et al. 1998). The use of trip-based SCFs can lead to reduced emissions estimates because the VMT distribution for a complete trip, at an overall average speed, is different from the VMT distribution for a particular link at the same overall average speed.

There are various other differences in the detailed implementation of EMFAC2000 and MOBILE6. These can lead to significant differences in the emissions estimates from the two models. However, the two models share the same overall approach and neither model provides any guidance for an innovative approach to the estimation of on-road mobile-source emissions.

MOBILE-SOURCE EMISSIONS MODELING IN THE FEDERAL REPUBLIC OF GERMANY

The Federal Republic of Germany has developed a set of mobile-source emissions models for a number of purposes (UBA/SAEFL 1999). There are three primary models that have different levels of complexity:

TABLE 5-1 Comparison of the Emissions Categories in MOBILE6 and EMFAC2000

MOBILE6				EMFAC2000			
	HC	CO	NO_x		HC	CO	NO_x
Normal	R £ 2	R £ 3	R £ 2	Normal	R £ 1	R £ 1	R £ 1
				Moderate	1 < R £ 2	1< R £ 2	1 < R £ 2
High	R > 2	R > 3	R > 2	High	2 < R £ 5	2 < R £ 6	2 < R £ 3
				Very High	5 < R £ 9	6 < R £ 10	3 < R £ 4
				Super	R > 9	R > 10	R > 4

Note: R is the certification standard. Emissions categories are defined in relation to R. For example, normal emitters in MOBILE6 have emissions of HC and NOx less than or equal to 2 times R.

- ***Handbook of Emissions Factors (Handbook)***[1]—This is a detailed model for calculating mobile-source emissions factors for a variety of driving conditions and vehicle types; the level of detail for this model is high compared with other German models. This model serves as the foundation of many of the emissions assessments made.
- ***CITAIR***—This model is used to predict emissions levels and pollutant concentrations for a variety of microscale control measures; essentially, this is a dispersion model combined with a set of emissions factors (not too dissimilar from CALINE4 or CAL3QHC).
- ***TREMOD***—This is more of an aggregate model that is used to calculate the total emissions inventory for the entire country based on emissions from the entire transportation sector (i.e., road traffic, rail, air, and ship) (IFEU 1997).

The three different models listed above are highly interrelated, with data files shared between the different models. For example, many of the emissions factors used in TREMOD come directly from the Handbook.

This set of emissions models is used for a number of purposes, including

- assisting in the development of standards of emissions protection;
- performing environmental assessment studies;
- road planning; and
- establishing permits for construction;

Over the years, the German mobile-source estimation techniques have become more and more refined (similar to the incremental improvements to the MOBILE series of models). It is expected in future years to be even more sophisticated. Of most relevance to the MOBILE model are the Handbook and TREMOD models, which are briefly described below.

Handbook of Emissions Factors

The Handbook of Emissions Factors (UBA/SAEFL 1999) is essentially a database program that is capable of accepting a number of user inputs, combining appropriate data sets, and predicting emissions factors for several situations. The Handbook is programmed in Microsoft ACCESS, a

[1]The Handbook was developed with participation from Switzerland and Austria; thus, it is used in all three countries.

flexible, programmable database application and is contained entirely within a single CD-ROM.

The Handbook is used to provide emissions factors for a variety of applications. It is often used in conjunction with traffic-simulation models in which emissions are estimated for different roadway sections in a transportation network. The Handbook also provides key emissions factors for the other emissions models, CITAIR and TREMOD.

The Handbook database contains information on several aspects of mobile-source emissions: (1) different vehicle categories, (2) different traffic compositions of those vehicle categories, (3) different traffic scenarios, (4) cold- and warm-start emissions factors, (5) evaporative emission factors, (6) different years of reference, (7) ambient temperature profiles, and (8) different functions for different species of emissions.

In creating the Handbook, two primary components were developed: an emissions-behavior component, and a driving-behavior component (see Figure 5–2). To characterize driving behavior, a number of instrumented vehicles were used to measure real-world driving patterns. A large set of velocity-time profiles were collected for a wide range of driving conditions, ranging from high-speed Autobahn conditions to stop-and-go traffic in urban centers.

Statistical analysis was then performed on the large driving-behavior data set, resulting in a number of representative driving cycles for different road types and different congestion conditions. All together, 43 representative driving cycles were created: 13 urban, 3 rural, 14 highway, and 13 special driving conditions.

In the parallel component, continuous (i.e., second-by-second) tailpipe emissions measurements were made during various chassis dynamometer tests. These dynamometer tests were applied to a wide variety of vehicle types, using several standard driving cycles (e.g., the FTP, the New European Driving cycle (NEDC), the U.S.-Highway cycle, and the German Autobahn cycles).

Approximately 300 LDVs were tested in the basic program, representing 15 different gasoline-fueled and 6 diesel-fueled vehicle types. The second-by-second emissions data for these different categories were matched with their corresponding instantaneous velocity and acceleration values, and velocity-acceleration indexed lookup tables were created to represent the emissions.[2] These lookup tables were filled out using interpolation techniques. For HDVs such as trucks, a similar methodology was applied, using the European transient cycle and steady-state emissions measurements made on engine dynamometers.

---

[2]This type of instantaneous or "modal" emissions modeling is described in more detail in a following section.

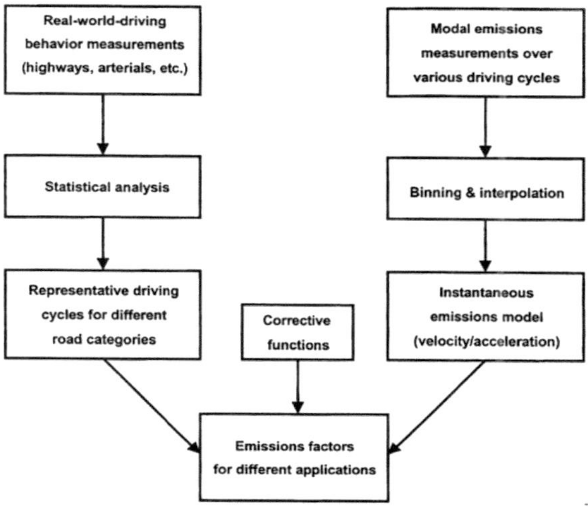

FIGURE 5–2 Database development for the Handbook of Emissions Factors. Source FRG-FEPA 1993.

The representative driving patterns derived as part of the driving-behavior component were then combined with the instantaneous emissions functions (i.e., lookup tables) representing the different vehicle types. For every second in a representative driving cycle, the emissions for a particular vehicle type can simply be looked up, and all the second-by-second emissions values for the specific driving cycle can then be summed together to represent an emissions factor.

In addition to the hot-stabilized emissions factors, supplementary testing was performed to provide additional correction factors for changes in road grade and for cold- and warm-start effects. By combining the driving-behavior database, the emissions-function database, and the added correction factors, emissions of CO, HC, NO_x, particulate matter (PM), CO_2, and a few other emissions species can be predicted.

Compared with the MOBILE model, Germany's Handbook of Emissions

Factors is somewhat more detailed and disaggregated in its emissions predictions. One of the key differences is that the emissions factors represented in the Handbook have been derived from the ground-up, using instantaneous emissions models developed specifically for producing emissions inventories for a wide variety of vehicles. MOBILE, in comparison, derives its emissions factors from integrated certification emissions testing, with additional correction factors for speed. Instead of using a global set of speed correction factors for all types of driving, the Handbook has also derived and established a wide range of representative driving cycles, something that MOBILE is now attempting to do in MOBILE6 with its facility and congestion cycles (see Chapter 3). Other key differences are that the German Handbook also has corrections for road grade, something MOBILE does not have.

TREMOD—"The Traffic Emissions Estimation Model"

In addition to the detailed Handbook, another macroscale emissions model was created for the entire transport sector of Germany. The TREMOD model was developed in 1993; it also uses the Microsoft ACCESS program. TREMOD was designed to compute emissions of CO, VOCs, NO_x, PM, and other species of emissions from all vehicles in Germany, including motor bikes, cars, trucks, airplanes, ships, buses, tractors, and trains. In addition, fuel consumption is also computed. The model is capable of predicting the transport sector emissions inventory for base years ranging from 1980 to 2020. The model uses extensive fleet characteristics and activity patterns (for past, present, and future years) for all transport modes.

TREMOD has been validated by comparing its overall predicted fuel consumption with collected fuel sales data. For gasoline, TREMOD predictions match very well. For diesel fuel, the match was not as good, primarily because diesel fuel is used in many different parts of the transport sector (e.g., military, agriculture, and stationary generators) where it is particularly difficult to estimate fuel consumption.

FUEL-BASED EMISSIONS INVENTORIES

The majority of regional emissions models in the United States, such as MOBILE and EMFAC, use travel-based models that combine gram-per-mile emissions factors with activity data in the form of VMT to estimate emissions. In contrast, fuel-based emissions inventories can also be calculated by normalizing emission factors to fuel consumption rather than VMT. Typically, fuel-based emissions factors are calculated from on-road emissions measurements (e.g., from remote sensors and tunnel studies).

The activity in this case is a measure of the amount of fuel consumed (Singer and Harley 1996). This methodology assumes that a precise fuel-use data set is readily available from records such as fuel taxes. Results of fuel-based emissions estimates are contained in the model evaluation section of Chapter 4. Here we will briefly describe the approach as well as some of its limitations.

In recent years, much has been learned about on-road vehicle emissions through the use of remote-sensing instruments. These instruments use an infrared source of light, and when the beam travels through an exhaust plume, it is possible to measure the spectral absorption. Measurements are made of the following ratios: CO to CO_2, VOC to CO_2, and in the newer sensors, NO to CO_2. With these measurements, it is possible to relate the amount of pollutant emitted to the amount of fuel burned using carbon-balance equations (Singer and Harley 1996). Further, since it is possible to obtain vehicle information (e.g., make, model, and vehicle type) by reading the license plate and applying it to a vehicle registration database, the fuel-based emission factors can be disaggregated within the vehicle fleet.

Vehicle activity is given by fuel-use data, which can be derived from tax records of fuel sales within each state. Spatial apportionment can be determined by tracking fuel shipments and performing filling station surveys. To determine the fuel-use activity of disaggregated vehicle subgroups, it is necessary to calculate the relative fuel economies between the subgroups and the travel fractions of the subgroups. The travel fractions can be determined by measuring the frequencies at which vehicles of each subgroup pass a remote sensor.

The accuracy of a fuel-based inventory depends highly on two factors:

- How well the entire vehicle fleet is represented by the remote-sensing measurements. Remote-sensing measurements are sensitive to a number of factors, including site location, speed and acceleration of vehicles, and road grade. The remote-sensing sites should be well distributed geographically within the area of study. In general, large numbers of measurements from each remote-sensing site are required to ensure that average emissions factors are determined accurately for all vehicle model years.
- How well the fuel-use activity data is accurately and correctly apportioned within the area of study.

In summary, the use of gram-per-gallon instead of gram-per-mile emissions factors is claimed to be a simpler method to calculate an emissions inventory, as long as sufficient remote-sensing and fuel-sales data is readily available. Many remote-sensing studies are taking place around the world and the use of remote-sensing in I/M programs are providing additional data. As newer remote sensors are used, VOCs and NO_x inventories might also be calculated.

It is important to point out that fuel-based methods are not designed to predict emissions inventories for future years, as does the MOBILE model. Fuel-based methods also do not provide spatially and temporally disaggregated emissions needed for air-quality modeling. In their present form, the fuel-based approach is useful as an independent method for verifying predictions from the traditional emissions inventory models, as shown in Chapter 4.

MODAL AND INSTANTANEOUS EMISSIONS MODELING

The MOBILE model (as well as CARB's EMFAC model) was developed for calculating regional emissions inventories using aggregated vehicle emissions data and estimates of vehicle activity in the form of VMT and average speed. Because of the inherent emissions and vehicle operation "averaging" that takes place in MOBILE, it is not suitable for evaluating traffic operational improvements that affect traffic and driving dynamics. For example, operational improvements that improve traffic flow (e.g., ramp metering, signal coordination, and automated highway systems) cannot be evaluated accurately with an aggregated model such as MOBILE.

The problem is that MOBILE uses average speed as the only variable for representing driving dynamics. Vehicle emissions are strongly coupled with driving dynamics, and average speed often does not properly characterize these dynamics. A large number of different driving patterns can have approximately the same average speed, but might have totally different driving dynamics and thus drastically different emissions responses.

To better capture emissions effects associated with a wide range of driving dynamics, researchers have investigated at a more fundamental level the modal operation of a vehicle and related emissions directly to vehicle operating modes such as idle, steady-state cruise, and levels of acceleration and deceleration. Models that can predict emissions based on these vehicle-operating modes are often referred to as modal emissions models. In general, several emissions modeling approaches have been introduced that attempt to include additional parameters beyond average speeds to better characterize emissions. The terms modal, instantaneous, and continuous are often used as synonyms when referring to this detailed microscale emissions modeling.

As described in previous chapters, MOBILE is based on emissions testing in which a single average emissions value is determined for a particular driving cycle. In contrast, modal or instantaneous emissions data collection consists of measuring emissions continuously during the chassis dynamometer tests and recording these data at a particular time interval, usually every second. Vehicle operational data are also recorded, such as the instantaneous vehicle speed and acceleration rate.

Speed-Acceleration Lookup Tables

The most basic and most common form of a modal or instantaneous emissions model is a multidimensional lookup table. Given one or more vehicle-operating variables, a table can simply store the corresponding emissions value. The most common emissions table is two dimensional, with the rows representing a velocity interval and the columns representing acceleration (see Figure 5–3). During an emissions test, all of the emissions measurements are put into different cells in the emissions matrix, according to the velocity and acceleration of the measured vehicle at that particular time. Some researchers use a "load" term (e.g., the speed-acceleration product) rather than acceleration for one of the table dimensions (Sturm et al. 1998). To guarantee the correct emissions value for every possible operating condition, a wide range of real-world driving cycles should be applied. However this is often impractical; therefore, a few driving cycles are applied, filling many cells in the emissions matrix. Values for the remaining cells are then interpolated from the data at hand.

The emissions lookup tables can be created for individual vehicles, or consist of a grouping of vehicles, based on common vehicle attributes (e.g., model year and technology type). When this form of an emissions model is used, an applied driving cycle (i.e., velocity-time profile) is considered one time step at a time, an emissions value is obtained from the lookup table, and all emissions values are then summed together to obtain an emissions value for the entire cycle.

There are several modal emissions models that are based on this lookup-table technique. In turn, there are also many traffic-simulation models that use this type of instantaneous emissions model. Work is currently being carried out by Oak Ridge National Laboratory (ORNL) to create speed-acceleration emissions lookup tables for Federal Highway Administration's (FHWA's) NETSIM traffic models (West and McGill 1997). ORNL uses a two-step process in which a vehicle is first driven through its entire operating envelope and its velocity is simultaneously measured second-by-second. The velocity pattern is then repeated while the vehicle is on a chassis dynamometer, measuring emissions to populate the emissions lookup tables. To date, ORNL has created modal emissions tables for 13 vehicle types. In Europe, the Handbook on Emissions Factors (BUWAL 1995) is used in Germany, Switzerland, and Austria, and its velocity-acceleration lookup tables are used for predicting instantaneous emissions (see earlier in this chapter). Joumard et al. (1995) use a similar method for calculating instantaneous emissions in France. Sturm et al. (1997) also have developed a model using this lookup-table technique.

The instantaneous emissions model based on lookup tables is a straightforward model to implement, and the computational costs are very low. However, there are several potential problems with this type of model.

First, it is crucial that a wide range of vehicle-operating conditions are used when developing the lookup tables, which might require a good deal of testing time. Second, when using instantaneous lookup tables, there is no explicit accounting for the time dependence in the emissions response to the vehicles operation. Many vehicle types exist for which vehicle-operating history (i.e., the last several seconds of vehicle operation) can play a significant role in an instantaneous emissions value (e.g., the use of a timer to delay command enrichment, and oxygen storage in the catalytic converter). If the instantaneous lookup tables were derived from statistical analysis of cycle-based data, the operating history effects could be considered to be inherently accounted for. However, this has yet to be validated. Third, there is no convenient way to introduce other load-producing effects on emissions such as road grade, or accessory use (e.g., air-conditioning), other than introducing numerous other lookup tables, or perhaps applying a set of corrections.

	DECELERATION/ACCELERATION (mph/s)												
Speed (mph)	-6	-5	-4	-3	-2	-1	0	1	2	3	4	5	6
0							IDLE						
5													
10													
15													
20													
25													
30							CRUISE						
35													
40													
45													
50													
55													
60													
65													

FIGURE 5–3 Speed and acceleration mix containing modes of idle, cruise, and different levels of acceleration and deceleration.

Aggregate Modal Emissions Models

Washington et al. (1997) describes the development of an aggregate modal emissions modeling approach using sophisticated statistical techniques. The model was developed by first analyzing in detail a large emissions certification database. Hierarchical tree-based regression analysis was then applied to the database, using several vehicle technology and operating characteristics as variables to explain emissions variations. Surrogate variables were also introduced as potential explanatory variables. The tree-based analysis searches for variables that explain the most variance in emissions response. For a set of vehicles tested over a variety of test cycles, the technique attempts to determine what variables have the greatest effect on overall emissions values. A regression tree is formed from the analysis, with the leaves of the tree providing grams/sec emissions rates for the specific mutually exclusive vehicle technology groups and operating characteristic combinations that naturally result from the regression-tree analysis (Washington et al. 1997). Both individual vehicle technology characteristics and operating mode characteristics appear in the tree. It was found that operating characteristics that had the most explanatory power were surrogate variables of acceleration conditions and power demand.

Like other methods, this modeling approach is limited by the representativeness of vehicles and cycles tested. Therefore, the greater the diversity in vehicles and emissions testing cycles, the more reliable the regression-tree model. Although more than 23,000 vehicle tests have been employed in this aggregate modal model development to date, there are too few recent model year vehicles represented in this database. Nevertheless, a strength in this approach is that the algorithms can be re-estimated on an annual basis, as new testing data become available on any number of vehicles and cycles.

The eventual form of this modal model will include hot-stabilized emissions rates and engine-start emissions rates. The model will also be capable of handling deterioration effects when the test age and odometer of the vehicle is included in the emissions database. This modal model is aggregate in the sense that it predicts a single integrated emissions value given any particular driving cycle. It does not provide instantaneous emissions values for every second of the driving cycle input.[3] This modal emissions

[3]It is important to note that is possible for this aggregate approach to predict at the second-by-second level using different explanatory variables, however the research team chose not to, in order to capture the largest amount of variability without over-complicating the model.

modeling technique is incorporated into Mobile Emission Assessment System for Urban and Regional Evaluation (MEASURE) modeling framework, described later in the chapter.

Neural-Network-Based Vehicle Emissions Models

Another approach uses a neural-network-based vehicle emissions model to simulate second-by-second emissions given an arbitrary driving cycle (Atkinson et al. 1998). This neural-network model is trained using dynamometer test results and makes nonlinear and multidimensional associations between vehicle-operating variables (i.e., speed and road load) and the emissions values. A particular neural-network architecture is first designed that allows accurate emissions prediction across the full envelope of vehicle operation. The network is then trained using a limited set of dynamometer-measured emissions values. The network "learns" the precise relationship between all designated inputs and outputs and can update those relationships over time to allow for engine wear, changes in fuel composition, or extreme combinations of operating conditions (Atkinson et al. 1998). This technique has thus far been successfully demonstrated on both light duty passenger vehicles and heavy-duty diesel vehicles. It can also be weighted to reflect the populations of the vehicle fleet when considering composite vehicles. Similar to the aggregate modal emissions technique described above, this modeling approach is limited by the representativeness of vehicles and cycles tested. Promising initial results have been documented. Given the extreme variability in vehicle sensors, control equipment, and deterioration factors, this modeling approach is not likely to provide a long term practical solution until a very large set of representative on-road data are available for such analyses.

Physical Instantaneous Emissions Models

Another approach to instantaneous emissions modeling is to use an analytical, physical modeling approach. In this type of approach, the entire emissions creation process is broken down into different components that correspond to physical phenomena associated with vehicle operation and emissions production (Barth et al. 1996). Each component is then modeled as an analytical representation consisting of various parameters that are characteristic of the process. These parameters typically vary according to the vehicle type, engine, and emissions technology. The majority of these parameters are stated as specifications by the vehicle manufacturers, and are readily available (e.g., vehicle mass, engine size, and aerodynamic drag coefficient). Other key parameters relating to vehicle operation and emissions production must be deduced from actual second-by-second emissions data.

This type of modeling is considered more deterministic rather than descriptive. Such a deterministic model is based on causal parameters or variables, rather than based on simply observing the effects (i.e., emissions) and assigning them to statistical bins (i.e., a descriptive model). This approach provides understanding, or explanation, for the variations in emissions among vehicles, types of driving, and other conditions. Using this type of model, analysts can gain insight to the physical and chemical reasons behind this model of emissions production.

This physical modal emissions modeling approach has been used in several models. Milkins and Watson (1983) were among the first to use this approach when developing emissions factors for vehicles in Australia. More recently, the Comprehensive Modal Emissions Model (CMEM) developed under sponsorship of the National Cooperative Highway Research Program (NCHRP Project 25–11) uses this approach (An et al. 1997; Barth et al.1996, 1997, 1998). Thus far, CMEM is capable of predicting engine-out emissions, tailpipe emissions, and fuel consumption for a comprehensive set of LDVs, in various states of condition (e.g., properly functioning, deteriorated, and malfunctioning). This model is based on a large, detailed database of second-by-second emissions data. Over 320 vehicles were tested to establish this model in which each vehicle underwent a comprehensive dynamometer testing procedure that consisted of a standard FTP test, the high-speed US06 cycle (to be used in future supplemental FTP testing, see Figure 3–4 and related discussion), and an in-house developed modal emissions cycle. This modal emissions cycle (MEC01) has been designed to include various levels of acceleration and deceleration, a set of constant speed cruises, speed-fluctuation driving, and constant power driving (Barth et al. 1997). CMEM has been validated against independent emissions measurements (albeit from the same vehicles used to create the model) and has shown good results. Additional validation efforts using independent vehicles and test conditions are currently in progress.

The physical modal emissions modeling approach has several attractive attributes:

- It inherently handles all of the factors in the vehicle-operating environment that affect emissions, such as vehicle technology, fuel type, operating modes, maintenance, accessory use, and road grade. Various components model the different processes in the vehicle related to emissions.
- It is applicable to all vehicle and technology types. When modeling a heterogeneous vehicle population, separate sets of parameters can be used

within the model to represent all vehicle and technology types. The total emissions outputs of the different classes can then be integrated with their correctly weighted proportions to create an entire emissions inventory.

- It is not restricted to pure steady-state emissions events, as is an emissions map approach, or a speed-acceleration matrix approach. Emissions events that are related to the transient operation of the vehicle can be appropriately modeled. Further, it can easily handle time dependence in the emissions response to the vehicle operation. As stated previously, the operating history (i.e., the last few seconds of vehicle operation) can play a significant role in an instantaneous emissions value.

Recent work by Jimenez (1999) also uses a physical-based approach for calculating an emissions inventory by investigating the relationship between emissions and vehicle-specific power (VSP). VSP is a vehicle's instantaneous power demand divided by its mass. VSP can be calculated by a number of physical parameters such as rolling resistance, aerodynamic drag, velocity, and acceleration. It is possible to develop a functional relationship between emissions and the single value of VSP, using data both from dynamometer measurements as well as remote-sensing measurements. Further, it is possible to generate an emissions inventory by creating a distribution of VSP using remote-sensing measurements (those that record instantaneous speed and acceleration) then multiplying this distribution by the precalculated VSP-emissions function. Preliminary results of this simplified method show promise.

A problem with both physical approaches described above is that there is tremendous variability in emissions within a vehicle class. Thus, to obtain an accurate estimate of both the mean and distribution of emissions from a particular vehicle type, a very large number of vehicles would have to be characterized.

INTEGRATION OF EMISSION MODELS WITH TRANSPORTATION MODELS

To calculate an emissions inventory, it is necessary to have both a vehicle activity component and an emissions-factor component. In Chapter 2, the methods of combining vehicle activity and MOBILE emissions factors for various applications have been briefly introduced. Earlier in this chapter, California's EMFAC model (also known as MVEI) is somewhat similar to MOBILE, except that the EMFAC modeling suite also includes a vehicle-activity component, known as BURDEN. This section focuses on the integration issues associated with combining vehicle activity and emissions factors across the various types of emissions-factor models.

In general, the integration of emissions-factor models and transpor tation-activity models or data can be represented as shown in Figure 5–4. Transportation models that produce vehicle-activity data are represented on the left, and the corresponding emissions-factor models/data are represented on the left. To produce an emissions inventory, output data from the transportation side is combined with appropriate factors from the emissions side.

Emissions factor models and their associated data and transportation models and their associated data vary in terms of their inherent temporal resolution, represented vertically in Figure 5–4. For example, at the lowest (microscale) level, traffic-simulation models typically produce second-by-second vehicle trajectories (i.e., location, speed, and acceleration). Driving cycles used for vehicle testing are also specified on a second-by-second basis (speed vs. time). At the highest (macroscale or regional) level, there are transportation models and data sets that aggregate with respect to time, producing traffic statistics such as average speed and total vehicle volume (i.e., VMT). At the midlevel (mesoscale), the transportation-activity results are more disaggregated than at the highest level, but still more aggregated than at the microscale level. For example, average speed might be provided not for the entire network, but rather on a roadway-facility basis. Other traffic dynamics statistics (i.e., average acceleration rates, and load) might also be provided at this mesoscale level.

In addition to temporal aggregation, vehicle aggregation must also be considered within the modeling category. At the most detailed level, unique emissions factors might be given for individual vehicles. However, it is unrealistic to model emissions at this level; what is typically done instead is to group similar types of vehicles together based on their class, technology, and model year. At the highest level of aggregation, vehicles might be categorized based simply on whether they are either passenger cars or trucks.

In the framework shown in Figure 5–4, the MOBILE model falls into the macroscale level of emissions factor models. When creating an emissions inventory, regional transportation models predict total vehicle volume and average speed for all vehicles (in many cases broken out on a facility-specific basis), which is then matched with the appropriate emissions factors from MOBILE. This is typically how a regional emissions inventory is produced.

However, when producing an emissions assessment of a traffic-flow improvement project or a specific intersection, it is more appropriate to perform modeling at a lower, more disaggregate level. In this case, microscale traffic-simulation models can produce second-by-second vehicle activity (in the form of velocity trajectories) that can be combined with modal (or instantaneous) emission models to produce an emissions inventory. Other applications at this level might estimate total emissions for a variety of

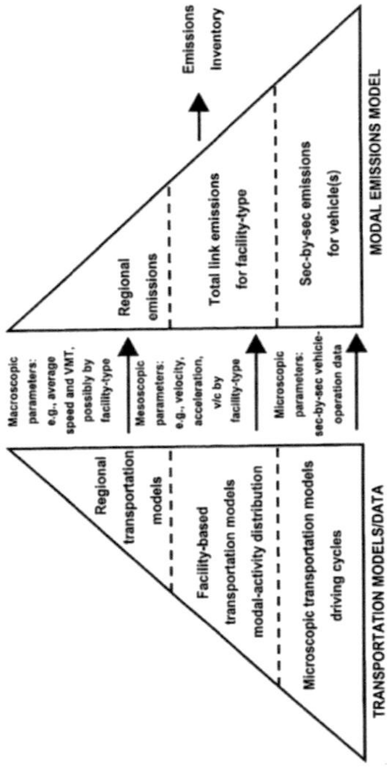

FIGURE 5-4 Transportation and emissions model interface.

vehicles given specific driving cycles, in lieu of performing expensive dynamometer tests. Also, vehicle-velocity patterns collected by instrumented vehicles, laser guns, and video-based computer vision can be directly input into an instantaneous emissions model to determine the total emissions associated with its activity.

As described in previous chapters, it is recognized that the conventional emissions models (e.g., MOBILE) have a number of limitations when producing a regional emissions inventory. As a result, version 6 of MOBILE is making a step in the right direction by disaggregating its representative driving patterns with its new facility and congestion cycles (see Chapter 3). Further, there have been several research efforts to develop models that produce regional emissions inventories at the mesoscale level. For example, the MEASURE model (described later) falls into this category.

It is important to point out that modal or instantaneous models can be used as the foundation for more aggregated emissions-factor models. An accurate instantaneous emissions model can be used to essentially replace expensive dynamometer testing. A driving cycle is simply applied, and the emissions associated with the cycle are produced. Therefore, with a modal emissions model providing the foundation, emissions factors can easily be created for models that have a wide variety of representative driving patterns. This is essentially what has been done with Germany's Handbook of Emissions Factors. Representative driving patterns were determined in a separate program from the emissions model component, and the emissions factors produced for these driving patterns were derived from a modal emissions model in the form of their velocity-acceleration-indexed lookup tables.

Further, SCFs used in MOBILE and EMFAC can also be improved with the use of a modal emissions model. SCFs have been created by performing emissions testing using a variety of driving cycles that have different average speeds. These emissions factors are then used to create the speed-correction curves as a function of average cycle velocity. When created these SCF functions in MOBILE and EMFAC, only a limited set of emissions testing has been carried out. With the use of a modal emissions model, many more factors could be produced for a wide range of driving cycles for many different vehicle types. Thus, the SCF functions would have a much stronger foundation, as long as a reasonably accurate modal emissions model was used in deriving them.

Microscale Traffic-Simulation Model Integration with Emissions Factors

At the microscale level of detail, traffic-simulation models can be combined with modal or instantaneous emissions models to predict emissions inventories. Second-by-second vehicle trajectory data are generated by the traffic-simulation model that can be used as input to the modal emissions model.[4] The resulting emissions data from all vehicles can then be integrated to provide a total emissions inventory. The easiest form of a modal emissions model to be applied here is the velocity-acceleration-indexed lookup table. In fact, the majority of microscale traffic-simulation models already have the built-in ability to predict emissions, given these emissions lookup tables. Keep in mind however, that the lookup-table form of model lacks the ability to handle road grade and vehicle operational history effects.

FHWA's TSIS suite of microscale models (i.e., FRESIM, NETSIM, and CORSIM) are capable of estimating emissions using this technique. In these models, the movement of individual vehicles are tracked on a second-by-second basis at intersections (NETSIM), corridors (CORSIM), and freeways (FRESIM) (FHWA 1998).

Other examples of microscale transportation models that operate in this fashion include the following:

- ***INTEGRATION***—A microscale traffic-simulation and dynamic-assignment model that traces movement of individual vehicles on freeways and arterials to a temporal resolution of 1 sec. Incorporating a built-in traffic-assignment algorithm, the model tracks the spatial and temporal activities of up to 500,000 vehicles operating on a subarea with a maximum of 10,000 links. INTEGRATION'S ability to combine arterial and freeway movements sets it apart from most conventional traffic-simulation models[5] (Van Aerde & Associates 1995).
- ***PARAMICS***—A suite of high-performance software tools for microscale traffic-simulation. Individual vehicles are modeled in fine detail for the duration of their entire trip, providing very accurate traffic flow, transit time and congestion information, as well as enabling the modeling of the interface between drivers and intelligent transportation system (ITS) technology. The Paramics software is portable and scalable, allowing a unified approach to traffic modeling across the whole spectrum of network sizes, from single junctions up to national networks. Key features of the Paramics model includes direct interfaces to macroscale data formats, sophisticated microscale car-following and lane-change algorithms, integrated routing functionality, direct interfaces to point-count traffic data,

[4]It is important to note that current traffic simulation models may not provide accurate vehicle speed/acceleration data (velocity vectors and/or speed/acceleration probability distributions) due to inadequate car-following equations. There is still a great deal of work that needs to be done in the traffic simulation arena.

[5]The name INTEGRATION comes from the model's ability to combine movements on arterials and freeways.

batch model operation for statistical studies, a comprehensive visualization environment, and integrated simulation of ITS technology elements (Paramics 1998).

Another microscale transportation model (in the sense that it tracks individual vehicles every second) is the TRANSIMS model, a large-scale program being developed under the sponsorship of the FHWA, EPA, and the U.S. Department of Energy. The details of this model and its emissions module are described later.

One of the key challenges for all of these microscale models is how to match the different vehicle types represented in the traffic-simulation component with the vehicle types represented within the emissions component. Traffic-simulation models typically have different vehicle types that are based on how they operate within a roadway network. In addition to the obvious divisions of vehicle types (i.e., motorcycles, passenger cars, buses, and heavy-duty trucks), categories are often made based on vehicle performance (e.g., high-performance cars and low-performance cars) that can be closely related to traffic-simulation parameters. For heavy-duty trucks, transportation models and data sets typically categorize their vehicles based on their configuration and number of axles. In all cases, a straightforward approach to handling the vehicle matching is to create an appropriate mapping between the vehicle types defined in the traffic-simulation model, and the vehicle types defined in the emissions model.

New Generation Research Transportation-Emissions Models

Microscale models track individual vehicles every second as they travel through a predefined roadway network. Because of the this detailed analysis, computer time and storage requirements can be high, depending on the size of the network. Therefore, a number of new generation research models are being developed that are not as aggregated as MOBILE, nor are they as detailed as the microscale models. These models are often referred to as "mesoscale models."

MEASURE

MEASURE is a model based on Geographic Information System (GIS) that uses an aggregate modal emissions model described earlier. The GIS framework allows for facility-level aggregations of microscale traffic-simulation, or disaggregation of traditional macroscale four-step travel-demand forecasting models to develop emissions-specific vehicle-activity data (Guensler et al. 1998).

The MEASURE model estimates both spatially and temporally vehicle activities that result in emissions. An emissions rate per unit of activity is defined for each of the activities. Several variables are addressed: vehicle parameters, operating conditions, fuel parameters, and environmental conditions.

The model is GIS-based to take advantage of the generation of spatial database management tools already being employed by state and metropolitan planning organizations for the management of municipal assets, resources, and activities. The GIS framework has been shown to be extremely versatile, allowing emissions estimates to be properly allocated spatially. Several key attributes of this model include the following (Guensler et al. 1998):

- ***Modular***—A modular approach has been taken so that individual model components can be independently assessed and validated.
- ***Stochastic***—Because of the high degree of variability in emissions, a stochastic modeling approach has been taken.
- ***Vehicle fleets***—Vehicle fleets are characterized by identifying distributions of different vehicle technology groups across space and time.
- ***Vehicle activity***—Both on-network activities and off-network activities are considered. Off-network activity (i.e., local roads) are handled on a zonal basis.
- ***Modal activities***—The model uses an aggregate modal modeling approach in combination with speed and acceleration distributions.
- ***Running-emissions rates***—Running-emissions rates are divided into two categories: hot-stabilized operation and enrichment conditions.
- ***Uncertainty***—An assessment of uncertainty is given with the model predictions.

The Integrated Transportation-Emissions Modeling (ITEM) Suite

In 1995, researchers began developing an integrated set of analytical tools that allows users to better assess the complex relationship between different traffic scenarios and emissions. This modeling suite is referred to as the Integrated Transportation-Emissions Model (ITEM) (Barth et al. 1995). ITEM was designed to incorporate highly time-resolved modal emissions data that are directly related to vehicle-operating modes, such as idle, various levels of acceleration and deceleration, and steady-state cruise. The modal emissions modeling component is the CMEM model described earlier.

ITEM'S transportation component is being developed on a hybrid macroscale and microscale approach. On the one hand, emissions data that are related to vehicle-operating characteristics such as acceleration and decel eration necessitate the detail found in microscale transportation models. On the other hand, a macroscale model is better suited for a large, regional traffic network. The computational requirements for a large, regional microscale model would be prohibitively high and would result in a model that is not very useful. By combining a macroscale traffic-assignment model with a set of microscale simulation models (organized by roadway facility type), both regional (i.e., wide-area network) and local (e.g., intersection) emissions inventories can be produced. Emissions are estimated as a function of vehicle congestion on particular roadway facilities, including freeway sections, arterials (with intersections), rural highways, and freeway on-ramps. Each microscale traffic-simulation model is tightly coupled with the macroscale traffic-assignment model, which can dynamically reroute traffic as network capacities change. A travel-demand model drives the traffic-assignment, thus a regional emissions inventory can be produced by using statistical emissions rates (as a function of roadway facility and congestion level) derived from the microscale components, and applying them to the individual links of the macroscale traffic-assignment model. The macroscale and microscale components are set up to run in parallel, so that users of the model can simulate real-time events (such as a traffic accident) and see the effect on traffic dynamics and emissions at both macroscale and microscale levels (Barth et al. 1995).

TRANSIMS

The Transportation Analysis SIMulation System (TRANSIMS) is a major effort aimed at fully integrating transportation and emissions models. TRANSIMS is being developed at the Los Alamos National Laboratory (LANL), funded by the U.S. Department of Transportation, FHWA, EPA, and the U.S. Department of Energy as part of the Travel Model Improvement Program. The overall goal is to deploy a large-scale transportation-simulation effort that integrates components of (LANL 1999)

- activity-based travel demand;
- intermodal trip planning;
- traffic microsimulation; and
- air-quality and other macro analyses.

The overall, unified architecture is shown in Figure 5–5.

The impetus for developing TRANSIMS stems from issues derived from the Intermodal Surface Transportation Efficiency Act, the CAAA90, and the introduction of various ITS implementations. New technical approaches are introduced in TRANSIMS to handle transportation-planning issues such as congestion pricing, alternative development patterns, transportation-control measures, and their effect on motor-vehicle emissions.

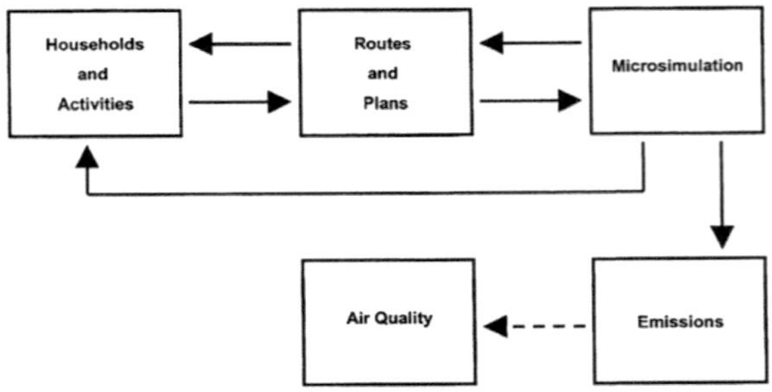

FIGURE 5–5 The four major modules of TRANSIMS include household and activity generation, intermodal router, traffic simulation and an emissions estimator. Note: Feedback loops are provided between the modules to replan and modify demand based on the results of traffic simulation. Source: LANL 1999.

TRANSIMS has several key features:

- The identity of individual synthetic travelers is maintained throughout the entire simulation and analysis architecture, with activity times and locations computed for each individual.
- The simulation output can provide a detailed, second-by-second history of every traveler in the system over a 24-hr day. Second-by-second dynamics of the traffic system can be observed in both local and global conditions.
- As illustrated in Figure 5–1, feedback paths are provided between modules in the simulation framework. These feedback paths provide stability in the results. Thus also allow for the simulation of various ITS strategies, such as simulating the movement of traffic information to selected travelers.
- TRANSIMS is highly modular. The individual modules can be replaced or modified without disturbing the overall TRANSIMS framework. Further, new modules can be introduced.

Framework

The flow among the different TRANSIMS modules is determined by a set of scripts. Intermediate data are collected in an iteration database to be used by other modules. In general the flow is summarized as follows:

- Given sufficient demographic data, synthetic household populations are created (at the desired level of detail) and distributed to match observed development patterns; typical demographic data include U.S. Census Bureau Public Use Microdata Samples and STF-3A data (data from the Census long form).
- Various activities for each traveler in the system (and freight movement) are generated. Activity patterns and mode-choice preferences are derived from surveys. Activity locations are determined based on standard gravity model methods.
- Individual travel plans are then produced for every individual and freight shipment. The intermodal planner computes a shortest or least-cost path for each traveler. The planner estimates the time that it takes to make a trip based on link traversal-time estimates contained in the overall network.
- Individual travel plans are then simulated on the network, on a second-by-second basis. The 1-sec update interval ensures that dynamic vehicle behavior is captured with a high degree of temporal fidelity.
- The environmental module then uses results of the microsimulation to predict tailpipe emissions for LDVs and HDVs. Evaporative emissions are also estimated. A total emissions inventory is produced and is used as input to various air-quality models (e.g., the MODELS-3 framework developed by EPA) to assess ambient concentrations of criteria pollutants at the regional or local level.

Environmental Module

The objective of the TRANSIMS environmental module is to translate vehicle behavior into consequent air-quality effects and energy consumption standards (Williams et al. 1999). Four major computational modules are required: emissions, atmospheric conditions, local transport and dispersion, and chemical reactions. The last three modules are handled using an air-quality model. The emissions module consists of

- an evaporation module, which treats emissions associated with resting losses, running losses, hot soaks, and diurnal pressure changes;
- an LDV emissions module, which includes aspects such as malfunc

tioning vehicles, emissions from cold- and warm-starts, normal driving, and off-cycle (i.e., non-normal) driving (when enrichment and enleanment events tend to occur); and
- an HDV emissions module, representing trucks and buses.

The evaporation module uses information from the microsimulation to determine the location of each vehicle and whether it is presently operating or has operated in the previous hour. If the vehicle has not been operating in the last hour, resting losses and diurnal evaporative emissions are calculated using the same formulation found in MOBILE6. While the vehicle is operating, running-loss emissions are calculated using the MOBILE5 formulation. If the vehicle has operated in the last hour, hot-soak start emissions are calculated based on the MOBILE6 formulation.

For LDV emissions, the comprehensive modal emissions model (discussed in an earlier section) is currently being integrated into the model. For the calculations, three sets of data need to be developed:

- ***Fleet composition***—This is determined from vehicle registration data, and I/M testing. Techniques have been developed to categorize vehicles into the appropriate CMEM category using vehicle registration information (Barth et al. 1998).
- ***Fleet status***—The status of each individual vehicle is tracked throughout the microsimulation. It is relatively straightforward to determine whether the vehicle is in a cold- or warm-start mode by simply tracking it through the network.
- ***Fleet dynamics***—One of the major challenges of the emissions module is to determine the dynamics of each vehicle as it is simulated in the traffic network.

The key problem is that the microsimulation component of TRANSIMS predicts second-by-second velocities at "quantum" steps, due to the cellular automata nature of the model. Each vehicle can occupy a 7.5 m spatial bin at any 1 sec; therefore, velocity can only assume one of several speed bins. To predict emissions due to vehicle dynamics (particularly during enrichment and enleanment events), the emissions module relies on additional empirical data of velocity-acceleration probability distributions (the MEASURE model described previously was a similar empirical approach). Using massive data sets from instrumented vehicles, cumulative distribution of accelerations have been derived as a function of the velocity-acceleration product. Three groups of acceleration are then determined: hard acceleration, insignificant acceleration, and hard deceleration. The acceleration rate for each vehicle is chosen based on the cumulative probability distribution. In addition, different roadway types and congestion levels are de termined from the microsimulation output, and using an additional empirical data set of typical velocity patterns for these roadway types and congestion levels, the fraction of the vehicles that undergo hard acceleration, insignificant acceleration, and hard deceleration are determined for the given context. The result is a continuous trajectory that can be fed into the modal emissions model to predict the emissions (Williams et al. 1999).

For HDVs, the fleet composition is broken down into buses and trucks. Various categories are then considered, based on engine size, chassis size, and model year. The fleet dynamics for the HDVs is less important with regard to emissions compared with LDVs. Buses and trucks typically have low accelerations and are usually driven at full throttle whenever the speed is less than desired and when there is adequate headway to accelerate. Thus, the modes of operation for heavy-duty-vehicles in TRANSIMS include full throttle, constant speed, and deceleration. The maximum acceleration is a function of engine size, road grade, and total vehicle weight. HDV emissions functions for TRANSIMS are derived from emissions testing performed at West Virginia University (Clark et al. 1999).

Status

A TRANSIMS deployment strategy has recently been developed to make the transition of TRANSIMS technology from a research and development project to a commercial product that can be used by transportation-planning agencies. The latest release of the TRANSIMS computer code is called TRANSIMS-LANL. LANL is currently seeking commercial developers for the code, and has released the code for evaluation purposes. The product commercialization process includes initiating licenses and contracts with vendors and developers to build product shells that package the TRANSIMS-LANL technology with user interface enhancements and other modules. TRANSIMS-LANL is also being released to various universities for research, development, evaluation, and demonstration purposes. It is expected that a commercial TRANSIMS product will be released by developers sometime in the year 2001.

SUMMARY

In summary, MOBILE is not the only motor-vehicle emissions model that exists. Various other vehicle emissions models have been or are being developed in other countries, at other regulatory agencies, and at different research organizations. These other modeling activities approach vehicle emissions estimation in a variety of ways. Some are very macroscale, similar to MOBILE. Others are much more microscale, looking at the vehicle emissions process at greater detail both temporally and spatially. One of the key points of this chapter is that some models are more appropriate in terms of their spatial and temporal resolution than others for a given application. It is clear that MOBILE cannot satisfy many of the applications that it is currently used for; therefore, the committee recommends consideration of an emissions modeling toolkit that incorporates a variety of emissions models for different applications. This is described in the next chapter.

It should be noted, though, that the upcoming version of MOBILE, known as MOBILE6, has undergone extensive peer review and now provides considerable documentation of the methodologies used in it. Any model that is used to replace MOBILE for specific applications must undergo a similar level of peer review and needs to provide in-depth documentation to any potential users. In addition, model validation must be the foundation for new model adoption. Validation efforts for all new modeling methods should be conducted with vehicles and test conditions not reflected in the data used to develop the model and be undertaken at the scale (or scales) for which a model is designed.

6

A Toolkit of Future Emissions Inventory Models

N THE PRECEDING chapters, the committee has taken a broad look at MOBILE. The report has discussed the uses of MOBILE in estimating mobile-source emissions, the technical issues associated with the model, issues associated with evaluating model uncertainties and accuracy, and alternative approaches for modeling mobile-source emissions. Two over-arching recommendations for improvements to the mobile-source emissions estimation process emerge from this review. The first is that the U.S. Environmental Protection Agency (EPA) should develop a long-term work plan, with input from the U.S. Department of Transportation (DOT) and others, on how to develop more accurate and effective mobile-source emissions modeling tools. The technical issues that should be addressed in this long-term plan are the focus of earlier chapters.

The second recommendation, the focus of this chapter, is that EPA should develop a modeling "toolkit" that better serves the full range of current uses of the MOBILE software. The motivation for this recommendation is that MOBILE is currently applied in situations for which it was not designed and is poorly suited. A "toolkit" of models is required, and here we lay out the structure for this proposed emissions modeling toolkit. The chapter proceeds with a discussion of the data needs and the user guidance that must accompany toolkit development. In closing, the chapter briefly touches on some of the institutional issues associated with the development and application of such a modeling toolkit.

REVIEW OF MOBILE'S USES AND SHORTCOMINGS

MOBILE was originally designed to estimate mobile-source emissions inventories and compare the effects of control strategies. As dictated by various legislative initiatives, including the Clean Air Act, Intermodal Surface Transportation Efficiency Act, and the National Environmental Policy Act (NEPA), an increasing number of requirements have been placed on planning organizations to better assess mobile-source emissions and overall air quality. As a result, MOBILE is now used for at least six general application areas (see Chapter 2):

1. National and regional regulatory strategies.
2. Evaluation of control strategies and emissions inventories, and rate of progress.
3. State Implementation Plans' (SIPs) demonstration of attainment.
4. Transportation conformity and evaluation of transportation impacts in a nonattainment area.
5. Transportation control-measure effectiveness.
6. NEPA and evaluation of capital investments.

These applications require assessment of mobile-source emissions at various temporal and spatial resolutions. Further, they often require interaction among the three different modeling disciplines: travel-demand, emissions, and air-quality modeling. Incremental improvements have been made to MOBILE over the years and MOBILE might still be well suited for aggregate regional and national analysis. However it cannot satisfy many of the applications listed above. Some of the individual issues associated with the use of MOBILE include the following:

- No protocol exists on the calibration of MOBILE model components.
- No protocol exists for standardizing emissions tests and ambient measurements made by the public and private sectors.
- No protocol exists regarding the evaluation of emissions-model estimates with air-quality measurements.
- No comprehensive assessment of MOBILE's sensitivity and sources of uncertainties has been completed that could help guide model improvements.
- Federal regulations regarding SIP time horizons (i.e., short term) are inconsistent with air-quality conformity rules calling for air-quality assessments from transportation plans and programs 20 years in the future. MOBILE, used as a regulatory tool, does not project technology impacts on the time scale required for conformity analysis very well.
- With significant decreases in mobile-source emissions rates for typical vehicles, off-cycle driving, high-emitting vehicles, and cold starts are

important sources of urban emissions. Little testing data, however, are available to estimate these emissions.

- Travel-demand, emissions, and air-chemistry modeling data sets and results are collected for separate purposes, yet these data and results must be used in a common framework to assess the air-quality effects of mobile-source emissions.
- With on-road mobile-source emissions decreasing nationwide, the contributions of non-road emissions to urban inventories are increasing substantially. EPA is giving inadequate attention to estimating and validating emissions from non-road sources.

These deficiencies and the wide variety of applications of MOBILE, demand a fresh look at setting priorities for work on mobile-source emissions modeling procedures. It is the opinion of this committee that MOBILE be supplemented with additional emissions modeling tools and data-collection efforts that will produce a better interface with transportation and air-quality models at the various levels of temporal and spatial resolution required.

DEVELOPMENT OF A MODELING TOOLKIT

The proposed emissions modeling toolkit should have several components, as shown in Figure 6–1. These are

- an *aggregated regional emissions-factor modeling component (i.e., the updated MOBILE model)* for estimating emissions using aggregate vehicle activity data;
- a *new-generation mesoscale emissions modeling component* that integrates detailed transportation and emissions components to estimate regional and subregional (corridor) emissions and air quality through the coupling of vehicle operating conditions with appropriate link-based or trip-based emissions factors;
- a *microscale instantaneous emissions modeling component* that uses instantaneous operating conditions of individual vehicles to estimate continuous vehicle emissions and can be used for a variety of applications, including generating emissions factors for microscale traffic-simulation models, mesoscale emissions models, traffic data sets, and dispersion models.

If implemented, the committee believes these tools will provide the necessary broad suite of models that are needed for sound policy-making. They will enable better assessment of the health and environmental consequences of mobile-source emissions-control programs.

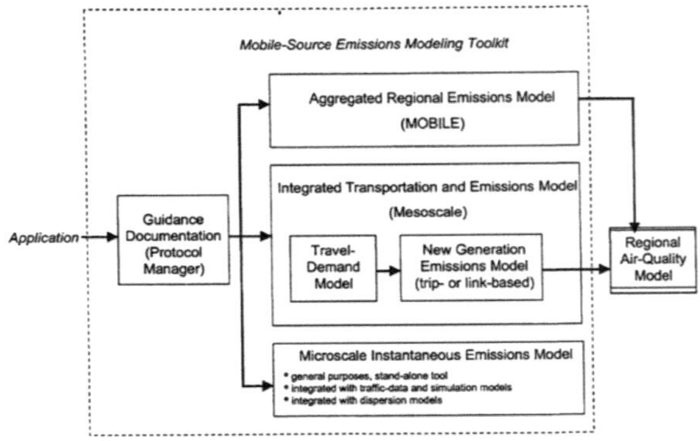

FIGURE 6–1 Schematic diagram of the mobile-source emissions modeling toolkit.

Regional Emissions-Factor Component

MOBILE6 and its preceding versions have been developed as an aggregate estimation method that works reasonably well for national and regional applications. MOBILE should remain as the aggregated regional emissions-factor component in the new suite of emissions models. This type of modeling component is required for comparisons of some control strategies and for comparing emissions from mobile sources with other source categories. It can estimate emissions inventories when provided with aggregated vehicle-activity data, as well as be used for evaluating new vehicle emissions standards, fuel specifications, and inspection and maintenance (I/M) program effectiveness. Further, it can be used to estimate the contribution of on-road vehicles to the nation's total pollutant emissions inventory and to monitor historical trends. This modeling component should be frequently upgraded.

Data Needs for the Regional Emissions-Factor Component

The data needs of the regional emissions-factor component of the modeling suite should not differ drastically from that which is required in cur rent MOBILE implementations. Typical user inputs include data such as vehicle technology, average speeds, ambient temperatures, fuel characteristics, tampering rates, fleet and vehicle miles traveled mix, mileage accumulation rates, registration distributions, basic exhaust-emissions rates, trip-length distributions, and operating-mode distributions. There are also assumptions about I/M program characteristics and credits and high emitters among others, as well as corrections for load, humidity, and air-conditioning effects. Metropolitan areas should be encouraged to develop their own input data for MOBILE, reducing the reliance on general default inputs that might not represent local conditions.

Mesoscale Transportation, Emissions, and Air-Quality Modeling Component

As discussed in Chapter 5, a new generation of mesoscale transportation and emissions models is currently under development. These models offer much better promise for satisfying many of the needs required in the six application areas. With this components ability to couple vehicle operating conditions to traffic flow, a mesoscale transportation, emissions, and air-quality model has the potential to assess emissions impacts of a wider variety of controls and conditions than the less-detailed regional emissions-factor component.

Calculating emissions inventories requires estimates of traffic flow and vehicle activity for different vehicle categories over the roadway network links. The mesoscale emissions modeling component would have a corresponding set of vehicle types and a set of emissions factors that correspond to different types of roadway facilities at different levels of congestion that can be applied on a link-by-link or trip-by-trip basis. These emissions factors can be established through comprehensive testing and the application of the microscale instantaneous emissions modeling component, described in the next section. Through the simulation of fleet composition, congestion, time-of-day, and other parameters, the model will produce a spatially and temporally resolved emissions inventory.

Pollutant concentrations for regional areas can be estimated for regional areas when the mesoscale model provides inputs to an air-quality model. This combination of models can then be used to demonstrate SIP attainment, such as for ozone, or used in determining whether major transportation plans conform to SIP requirements. An additional application at the mesoscale level is a detailed evaluation of I/M effectiveness. The mesoscale emissions modeling component should have sufficiently detailed temporal (hourly estimates) and spatial (1–5 km gridded) allocations that match well with current regional air-quality models. Because of the need for a high level of spatial detail, it is strongly recommended that this mesoscale model component be developed within a Geographical Information System (GIS) environment.

The mesoscale emissions modeling component should predict not just emissions inventories of nitrogen oxides (NO_x), volatile organic compounds (VOCs), and carbon monoxide (CO), but also fuel consumption, particulate matter (PM) and air toxics emissions, and greenhouse gas emissions (especially carbon dioxide (CO_2)). As with the MOBILE model, the committee sees no need for separate models for different pollutants (e.g., a model such as PART separate from MOBILE to simulate PM emissions). Given an estimate of fuel consumption, the estimation of CO_2 and some other greenhouse gases is straightforward.

As described in Chapter 5, TRANSIMS is a major transportation, emissions, and air-quality modeling framework that has the potential to fill a portion of the requirements recommended for the mesoscale modeling component. TRANSIMS is an integrated system of travel-demand and emissions models simulating a detailed representation of a given population's travel behavior and the resulting emissions. However, the model will not be fully developed for several years. For this model, or any model, to be used in regulatory applications, it must undergo peer review such as the MOBILE model currently does. It also must be extensively validated and documented for users. TRANSIMS might fulfill this regulatory role; other new-generation models, such as the MEASURE and ITEM models described in Chapter 5, might also support this integrated transportation and emissions component in the toolkit. It cannot be overemphasized, though, that any model intended to fulfill such a regulatory role must undergo extensive peer review and validation, and provide in-depth documentation to any potential users.

Data Needs for Mesoscale Transportation and Emissions Modeling Component

To estimate emissions inventories more accurately using the integrated mesoscale modeling component described above, will require a greater array of input data. These input data also require more spatial and temporal resolution. This implies a need for more detailed data from state and local agencies. Some states and other users may find it difficult to develop the detailed activity-level data required to implement these models. However, many agencies have already begun to build detailed local data sets, some of which are represented in a GIS framework.

Transportation modeling requires better data in a number of areas (Chatterjee et al. 1997):

- ***Demographic data***—These data are important for determining trip-generation factors and include population by age, gender, and density;

household size and number of dependents; and projections of these demographic variables in future years.

- ***Economic data***—Important economic variables must be identified and developed, including number of households, income and vehicle ownership by household, employment, and projected future growth.
- ***Activity data***—These data help define the present and future number, purpose, duration, mode, and other parameters for trips made by households.
- ***Land-use data***—These data are used for determining the effect of planning and zoning on transportation system utilization and performance. These data include residential and employment land-use fractions, concentrations of residential and employment land use, access from residential and nonresidential areas to transit stops, and future land-use projections.
- ***Roadway link data***—To effectively model the road network, data such as road segment length, number of lanes, posted speed, link capacity, and road-facility type, are needed.
- ***Transit data***—Both nonrail (i.e., buses and paratransit services) and rail transit data are required, such as round-trip travel times, average speed and stop times, stop locations, peak and off-peak service frequencies, and direction of service.
- ***Regional measures***—These data include production-attraction counts at zonal levels, VMT by road-facility class, mode splits, and preference surveys for transit and HOV lane usage.
- ***Microscopic measures***—Link level volume-to-capacity ratios, average speed, travel times, transit ridership by hour, and percent of truck usage.

Likewise, the emissions module will also require more detailed input data. It will require more spatial and temporal disaggregation of the same variables that are included in MOBILE including link-based emissions factors for different speeds and congestion levels, start distributions, and disaggregated rates for hot-soak, diurnal, resting-loss, and running-loss rates. This module will also require more detail on the vehicle fleet, such as registration distributions by vehicle class, fuel type, and emitter-level category.

Microscale Transportation and Emissions Modeling Component

A critical component to the emissions modeling toolkit is a microscale instantaneous emissions model. As described in Chapter 5, an instantaneous or modal emissions model predicts emissions for a variety of different driving dynamics and can be used for a variety of applications.

- It can be combined directly with microscale traffic-simulation models (or measured data) to provide intersection and corridor-level emissions inventories.
- It can be used to reduce expensive dynamometer emissions testing.
- It can serve as the foundation for other components in the modeling toolkit. For example, given a set of roadway facility and congestion cycles, it can be used to determine link-based emissions factors; further, generalized speed-correction factors (SCFs) used in the regional aggregate emissions-factor component can be improved with an accurate instantaneous emissions model.

Some of the applications that require an accurate coupling of transportation and emissions modeling components can be performed using a combined microscale transportation and emissions model set. This is an area in which MOBILE has major problems. A microscale emissions model can be used to evaluate transportation control-measure effectiveness and some Congestion Mitigation Air Quality projects. Microscale transportation, emissions, and dispersion models also can be combined to determine pollutant concentrations for a variety of transportation projects.

Data Needs for the Microscale Transportation and Emissions Modeling Component

As described in Chapter 5, several instantaneous (modal) emissions models are being developed to predict second-by-second tailpipe emissions for a variety of vehicle and technology categories. The development of these models requires extensive vehicle emissions testing, with an emphasis on real-world vehicle operation outside the performance envelope of the Federal Test Procedure (FTP). This detailed vehicle emissions testing should continue on a yearly basis to capture the effects of changes in fuel and automotive technology. Although instantaneous emissions models for light-duty vehicles (LDVs) are now becoming available, much more emissions data will be required to construct a comprehensive model for heavy-duty vehicles (HDVs). Another important component is determining the fraction of high-emitter in the on-road vehicle fleet as well as high-emitter emissions rates. Cooperative public-private partnerships in data collection are critical to expedite the development of necessary information.

When the microscale emissions and transportation modeling component is combined with detailed traffic-simulation models, the models have similar data needs as specified above, with additional data requirements such as signal types, intersignal spacings, signal phasing, mean start up delay at each signal, upstream distance of freeway exit signs from exits, and emergency-response vehicle times.

Differences Among Emissions Modeling Components

There are basic differences in the processes represented in the macroscale, mesoscale, and microscale emissions modeling components. These differences affect how well each is suited for various applications.

The macroscale emissions modeling component is the most aggregate in terms of coupling emissions with the operating modes and activity of vehicles. It uses parameters such as average vehicle speed, technology class and age, and VMT to estimate emissions for broad vehicle classes. The comprehensive nature of the model, representing emissions-causing processes for large numbers of vehicles, makes it useful for estimating large-scale emissions inventories (regional to national). However, the model does not directly link emissions to operating parameters that have a large impact on emissions, such as vehicle and traffic dynamics, and subsumes information about vehicle activity into parameters such as VMT and SCFs (see Chapter 3).

The mesoscale emissions modeling component contains a greater level of coupling between vehicle operating modes (i.e., vehicle activity) and emissions. The mesoscale level still uses "emission factors," but these are more closely coupled with travel-demand and traffic-simulation models to better represent the dynamic effects of traffic activity on emissions. Emissions can be associated with different kinds of driving, disaggregated for example by roadway facility type and by congestion level. There might need to be 50 to 100 cases of these mesoscale emissions factors that are indexed by transportation model links and different levels of activity. This is far more detailed than MOBILE5, in which a single set of SCFs are applied universally to all VMT, or MOBILE6, in which SCFs will be applied to four facility types at varying congestion levels. The mesoscale component should provide a greater level of spatial and temporal disaggregation of emissions and incorporate a larger range of parameters that are known to affect emissions.

The microscale emissions modeling component couples emissions with the instantaneous operating conditions of individual vehicles to produce a continuous (typically second-by-second) estimate of vehicle emissions. As such, it inherently handles emissions effects related to *vehicle* dynamics, and when coupled with microscale traffic-simulation models, it predicts emissions related to *traffic* dynamics. The emissions from individual vehicles are summed to estimate total emissions for a particular traffic situation during a particular time period. Because of the computational burden (and potentially large data-storage requirements), it is possible to do this only for relatively small scales, such as for intersections and corridors. Doing it for large regional areas is not feasible at this time. However, the instantaneous emissions modeling component can also be used to derive the mesoscale emissions factors. The microscale emissions model could essentially "precompute" the emissions factors by feeding in different representative driving cycles for the different links and congestion levels.

Consistency Among Models in the Toolkit

To allow for a nested progression of models, the models in the toolkit should be designed and maintained to be consistent, despite differences in the level of detail in inputs (e.g., variations in roadway network detail and meteorological variables) and the outputs (e.g., variations in emissions inputs over space and time). It is important that the different components in the proposed modeling toolkit are, to the best extent possible, based on the same data set and are able to predict similar emissions for similar conditions. A prominent example of how things should be consistent is that an instantaneous emissions model should generate the same integrated emissions numbers as an aggregate model for a specific driving cycle. Further, link-based or trip-based emissions factors can be generated directly from an instantaneous component, insuring consistency between those two layers. In general, the results of various emissions inventory methodologies should be as consistent as possible. When they do differ significantly, the modeling application should be evaluated in detail to determine the reasons for the differences, and changes in the models should be considered to reduce these differences.

GUIDANCE DOCUMENTATION

Guidance documents, suitable for the full range of expected users, are critically important and must be developed in concert with this modeling toolkit. These documents need to be specific about preferred methods and protocols in estimating emissions inventories for each of the six application areas. The guidance documents should not be developed solely by EPA, but rather it should include groups representing transportation, emissions, and air-quality disciplines. The MOBILE software has evolved over time in a rather haphazard way, seemingly backing into expanded uses in public-policy matters. The purpose of developing guidance or protocol management in an interdisciplinary setting is to allow the development phase to be comprehensive, open, strategic, and responsive. It is a core requirement for establishing trust in public-policy discussion as well as for ensuring broad and correct use of the models.

Changes in modeling must be made in response to the demand users place on the current paradigm. The modeling toolkit approach presented in Figure 6–1 is specifically designed to expand to each of the applications. The guidance documents and established protocol management define how the emissions modeling toolkit should be applied for any specific application. Examples of the six general applications areas follow:

Example 1: *National and regional regulatory strategies.* Work under this requirement would continue to be performed using MOBILE in the traditional manner.

Example 2: *Evaluation of control strategies, emissions inventories, and rate of progress.* These evaluation requirements can be performed in a number of different ways, depending on the inherent level of detail needed. For example, in smaller communities that do not have overly sophisticated transportation systems, the traditional aggregated, regional emissions model component (i.e., MOBILE) coupled with a traditional travel-demand model will likely suffice. Communities that have complex transportation systems would apply the mesoscale transportation and emissions technique (see Figure 6–1).

Example 3: *SIP demonstration of attainment.* As in example 2, this work could be performed at different levels of detail.

Example 4: *Transportation conformity and evaluation of transportation impacts in a nonattainment area.* Again, as in example 2, the same two approaches would be used to conduct these tasks. It is likely that large urban regions might opt for an integrated approach, whereas mid-sized urban areas might use stand-alone components of the modeling toolkit.

Example 5: *Transportation control-measure effectiveness.* Most transportation control measure effects are seen primarily at the microscale level-of-detail. Therefore, the microscale components of the toolkit are to be used specifically for this purpose. The microscale components in this case would consist of a traffic simulation model (and/or off-model statistical evaluations of modal activity changes) tightly coupled with the instantaneous emissions model component.

Example 6: *NEPA and evaluation of capital investments.* Again, the dual approach described in example 2 would be used; however, subarea values representing corridor conditions would have to be input into the traditional software.

Thus, the modeling toolkit will add flexibility in responding to the six application areas. It also permits a choice of approaches to address the more extensive application areas, permitting some regions and urban areas to use more integrated approaches, while others use a sequential, more aggregate approach to respond to air-quality requirements.

SUMMARY OF POLICY AND INSTITUTIONAL ISSUES

Many policy and institutional issues are associated with a toolkit ap proach to emissions modeling. MOBILE plays a leading role in mobile-source regulation and, by extension, in air-quality regulation. The development of a modeling toolkit approach will require a major initiative championed by a broad constituency. The committee feels strongly that such an effort is warranted, given the significance of mobile-source regulation. It should be emphasized, however, that this requires a very broad and substantial effort including coordination of planning, data collection, model development, documentation, and evaluation. EPA alone cannot develop an effective toolkit, and should not try to do so.

There are many potential partners for EPA in this effort. Both the California Air Resources Board (CARB) and DOT have the modeling interest and expertise as well as the regulatory perspective. The committee recommends that within 1 year of this report's publication these entities initiate a cooperative program to develop a suite of models for assessing on-road and off-road mobile-source emissions. This program should begin with an assessment of the modeling needs that the regulatory and scientific communities foresee in the next decade. Discussions should include LDVs and HDVs, and should address all criteria pollutants, toxics, PM-2.5, and greenhouse gas emissions. Following the assessment of modeling needs, these agencies and other interested parties should assess the developmental status of models that might be used within the toolkit. The product of this evaluation should be a plan for coordinating development of modeling elements. Finally, there should be an assessment of present and future data requirements. The emphasis should be on coordinating and standardizing data-collection efforts and developing partnerships with universities, industry, and national laboratories to aid in this effort.

The discussion of the need for a new suite of models, including a revised MOBILE model, raises the issue of creating a single regional emissions-factor model for use by all states. California has its own legal authority to set automotive vehicle emissions standards and its own ambient air-quality standards. Because of this, CARB has developed the EMFAC model, which is tailored to the automotive technology, fuels, and driving patterns in California (see Chapter 5). As with MOBILE, EMFAC is updated periodically. The most recent version, EMFAC2000 shows substantially higher estimates of emissions than the previous version of the model, EMFAC7G. With a gradual but perceptible closing of the gap between federal and California emissions standards, and hence automotive emissions-control technologies, the opportunity might arise in the future to further improve coordination of the EPA and CARB programs or even to combine MOBILE and EMFAC into one model. It is recommended that EPA and CARB immediately start to explore this possibility and to develop a time frame that is scientifically and technically appropriate. A joint report should be issued within a year of this report.

Another critical institutional issue concerns model evaluation and vali dation. The committee recommends that EPA and the other agencies involved in the development of a modeling toolkit not only undertake such studies, but also develop an institutional framework to ensure high-quality, independent studies are conducted. The committee does not recommend any particular institutional arrangement, but several possibilities available. For example, in the past EPA has relied on blue ribbon panels and committees organized under the Federal Advisory Committee Act, such as the Mobile Source Technical Review Subcommittee discussed in Chapter 3, to provide advice on critical issues. Another example is the Emission Inventory Improvement Program (EIIP), which is a cooperative effort among state and local agencies, EPA, and industry. It was started by the State and Territorial Air Pollution Program Administrators and the Association of Local Air Pollution Control Officials to develop procedures for collecting, estimating, and reporting emissions data. A similar organization could also be developed to coordinate model evaluation studies.

A final issue involves the need for support for a modeling toolkit initiative from the legislative community that is involved in setting mobile-source regulations but not in implementing and evaluating these regulations. There is a constant demand from this group for the quantification of emissions and air-quality impacts of mobile-source emissions-control programs. Models are critical in this effort because they provide a consistent framework for evaluation and because they reduce the need to develop evaluation methods on a case-by-case basis. Quantifying the impacts of some initiatives, however, are beyond current modeling capabilities. This is especially true for programs that have small emissions impacts, little support data, or large collateral consequences. Too often, legislators demand assessments of controls with a high level of accuracy in situations where insufficient data and evaluations methodologies exist. And, too often, regulators are unable to fulfill these requests. The difficulty of quantifying effects is often not appreciated by those who request evaluations. It is critical for legislators to understand the limitation of current modeling capabilities and where additional resources are needed to further development of these capabilities. Improving this situation requires improved communications between those who perform evaluations of control programs and those who mandate such evaluations.

References

An, F., M.Barth, and J.Norbeck. 1997. Development of a comprehensive modal emissions model: Operating under hot-stabilized conditions. Transportation Research Record 1587:52–62.

AQIRP (Air Quality Improvement Research Program). 1992. Effects of Fuel Sulfur on Mass Exhaust Emissions, Air Toxics, and Reactivity. (AQIRP Data Analysis and Report Writing Subcommittee). Technical Bulletin No. 8. Coordinating Research Council, Atlanta, GA.

Atkinson, C.M., N.N.Clark, T.L.Long and E.L.Hansevack. 1998. Neural network-based vehicle emissions modeling for inventory applications. An update on virtual sensing. Pp. 2–1–2–15. Proceedings of the 8th CRC Road Vehicle Emissions Workshop, San Diego, CA. April.

Austin, T.C., and T.R.Carlson. 1997. Development of Speed Correction Cycles. SR97–04–01. Prepared by Sierra Research, Inc., for the U.S. Environmental Protection Agency. April 30, 1997.

Barth, M., F.An, J.Norbeck, and M.Ross. 1996. Modal emissions modeling: A physical approach. Transportation Research Record 1520:81–88.

Barth, M., R.R.Tadi and J.Norbeck. 1995. The development of an integrated transportation/emissions model for estimating emission inventories. Pp. 1225–1236 in Transportation Congress: Civil Engineers-Key to the World Infrastructure, B.K.Lall and D.L.Jones, eds. New York: American Society of Civil Engineers.

Barth, M., T.Younglove, T.Wenzel, G.Scora, F.An, M.Ross, and J. Norbeck. 1997. Analysis of modal emissions from diverse in-use vehicle fleet. Transportation Research Record 1587:73–84.

Barth, M., F.An, T.Younglove, C.Levine, G.Scora, M.Ross and T. Wenzel. 1998. The Development of a Comprehensive Modal Emissions Model. Draft final report to the National Cooperative Highway Research Program.

Bauer, H. 1996. Automotive Handbook, 4th Ed. Stuttgart: Robert Bosch.

Bishop, G.A., and D.H.Stedman. 1996. Motor vehicle emissions variability. J. Air Waste Manage. Assoc. 46(7):667–675.

Brachaczek, W.W. and W.R.Pierson. 1974. Airborne particulate debris from rubber tires. Rubber Chemistry and Technology. 47(5):1275–1299.

BUWAL (Bundesamt fur Umwelt, Wald und Landschaft-Swiss Federal Office for the Environment). 1995. Luftschadstoff-Emissionen des Strassenverkehrs 1950–2010. SRU-255-D. BUWAL.

Cadle, S.H., P.Mulawa, E.C.Hunsanger, K.Nelson and R.A.Ragazzi. 1998. Measurements of Exhaust Particulate Matter Emissions from In-Use Light-Duty Motor Vehicles in the Denver, Colorado Area, NTIS/PB98–136401. Coordinating Research Council, Atlanta, GA.

Cadle, S.H., P.Mulawa, P.Groblicki, C.Laroo, R.A.Ragazzi, K.Nelson, G. Gallagher and B.Zielinska. 1999. In-Use Light-Duty Gasoline Vehicle Particulate Matter Emissions on the FTP, REP05, and UC Cycles. Final report. CRC Project E-46. Prepared for Coordinating Research Council, Atlanta, GA. June, 2.

CARB (California Air Resources Board). 1996a. Derivation of Emission and Correction Factors for EMFAC7G. California Air Resources Board, Mobile Source Control Division, El Monte, CA.

CARB (California Air Resources Board). 1996b. Methodology for Estimating Emissions from On-road Motor Vehicles. Technical Support Division, Mobile Source Inventory Branch, in seven volumes. October-November.

CARB (California Air Resources Board). 1998. Measuring Concentrations of Selected Air Pollutants Inside California Vehicles. Final Report. Prepared by Research Triangle Institute, AQMD Contract 98055, RTI Task 93U-6786–000 for California Environmental Protection Agency, Air Resources Board, Sacramento, CA. Dec. 28.

Carlock, M. 1999. An Overview of California's Emissions Inventory Model EMFAC99. Presentation to NRC Committee to review the MOBILE model, March 4, 1999.

Cha, S., P.Carter and R.L.Bradow. 1983. Simulation of Automobile Brake Wear Dynamics and Estimation of Emissions. SAE Technical Paper 831036. Society of Automotive Engineers, Warrendale, PA.

Chameides, W.L., R.D.Saylor, and E.B.Cowling. 1997. Ozone pollution in the rural United States and the new NAAQS. Science 276:916.

Chatterjee, A., T.Miller, J.Philpot, T.Wholley, R.Guensler, D.Hartgen, R.Margiotta, and P.Stopher. 1997. Improving Transportation Data for Mobile-Source Emissions Estimates. NCHRP Report 394. Washington, D.C.: National Academy Press.

Chico, T., M.E.Bassett, J.C.Cassmassi, S.Mitsutomi, X.Zhang, and H. Hogo. 1993. Application of the Urban Airshed Model for a SCAQS Episode in the South Coast Air Basin. Pp. 257–263 in: Southern California Air Quality Study. Data Analysis. Proceedings of the International Specialty Conference. Pittsburgh, PA: Air & Waste Management Association.

Chrysler Corporation. 1998. Emission and Fuel Economy Regulations. Environmental & Energy Planning. Chrysler Corporation, CIMS 482– 00–71. Auburn Hills, MI.

Clark, N.N, R.D.Nine, M.Gautam and C.M.Atkinson. 1999. Effect of Test Cycles on Measured Emissions of Diesel Vehicles. Proceedings of the 9th CRC On-Road Vehicle Emissions Workshop, San Diego, CA. April.

Clean Air Report. 1999. States may lose half of current I/M emissions credit in new air model. (April, 15):7.

Darlington, T., J.G.Heiken, and D.Kahlbaum. 1999. Nationwide and State-by-State emissions benefits of a gasoline sulfur limit. Pp. 578– 596. In: Emission Inventory: Living in a Global Environment. Vol. 1. Sewickley, PA: Air & Waste Management.

Davis, S.C. 1997. Transportation Energy Data Book: Edition 17. ORNL-6919. Center for Transportation Analysis, Oak Ridge National Laboratory, Oak Ridge, TN.

Davis, S.C. 1999. Transportation Energy Data Book: Edition 19. ORNL-6958. Center for Transportation Analysis, Oak Ridge National Laboratory, Oak Ridge, TN.

Dockery, D.W., C.A.Pope, X.Xu, J.D.Spengler, J.H.Ware, M.E.Fay, B.G. Ferris, Jr., and F.E.Speizer. 1993. An association between air pollution and mortality in six U.S. cities. N. Engl. J. Med. 329(24):1753– 1759.

Energy Information Administration. 1997. Emissions of Greenhouse Gases in the United States 1996. Washington, DC: United States Department of Energy.

Energy Information Administration. 1998. Annual Energy Outlook 1999 with Projections to 2020. DOE/EIA-0383(99). Office of Integrated Analysis and Forecasting. Washington, DC.: U.S. Department of Energy. December.

ENVIRON International Corporation. 1998. User's Guide for the National Nonroad Emissions Model. Draft version. Prepared for U.S. Environmental Protection Agency. National Vehicle and Fuel Emissions Laboratory. Ann Arbor, MI. June.

EPA (U.S. Environmental Protection Agency). 1986. Air Quality Criteria for Ozone and Other Photochemical Oxidants, Vol. II. EPA-600/8- 84/020bF. Environmental Criteria and Assessment Office, U.S. Environmental Protection Agency, Research Triangle Park, N.C. August.

EPA (Environmental Protection Agency). 1993a. Criteria and Procedures for Determining Conformity to State or Federal Implementation Plans of Transportation Plans, Programs, and Projects Funded or Approved Under Title 23 U.S.C. or the Federal Transit Act. Fed. Regist. 58(225): 62188. November, 24.

EPA (U.S. Environmental Protection Agency). 1993b. Final Regulatory Impact Analysis for Reformulated Gasoline. EPA420-F-93–017. Office of Mobile Sources, U.S. Environmental Protection Agency, Ann Arbor. December.

EPA (U.S. Environmental Protection Agency). 1994. User's Guide to MOBILE5 (Mobile Source Emission Factor Model). EPA-AA-TEB-94– 01. Office of Mobile Sources, Office of Air and Radiation. Ann Arbor, MI. May.

EPA (U.S. Environmental Protection Agency). 1995. Draft User's Guide to PART5: A Program for Calculating Particle Emissions from Motor Vehicles. EPA-AA-AQAB-94–2. Office of Mobile Sources. U.S. Environmental Protection Agency. Ann Arbor, MI. February.

EPA (Environmental Protection Agency). 1997a. Finding of Significant Contribution and Rulemaking for Certain States in the Ozone Transport Assessment Group Region for Purposes of Reducing Regional Transport of Ozone. Proposed Rule, Fed. Regist. 62(216): 60318. November 7.

EPA (U.S. Environmental Protection Agency). 1997b. Control of Emission of Air Pollution from Highway Heavy-Duty Engines. Final Rule. Fed. Regist. 62 (203):54694. October 21.

EPA (U.S. Environmental Protection Agency). 1997c. Analysis of the Arizona IM240 Test Program and Comparison with the TECH5 Model. EPA 420-R-97–001. U.S. Office of Mobile Sources. U.S. Environmental Protection Agency. Ann Arbor, MI.

EPA (Environmental Protection Agency). 1998a. National Air Pollutant Emission Trends Update: 1970–1997. EPA 454/E-98–007. Office of Air Quality Planning and Standards, U.S. Environmental Protection Agency, Research Triangle Park, NC. Available: http://www.epa.gov/ttn/chief/trends97/emtrnd.html

EPA (Environmental Protection Agency). 1998b. National Air Quality and Emissions Trends Report, 1997. EPA-454/R-98–016. Office of Air Quality Planning and Standards, U.S. Environmental Protection Agency, Research Triangle Park, NC.

EPA (Environmental Protection Agency). 1998c. Exhaust Emissions Certification Standards for Light Duty Vehicles and Light Duty Trucks.

EPA 420/B-98–001. Office of Mobile Sources, U.S. Environmental Protection Agency, Ann Arbor, MI. Available: http://www.epa.gov/oms/stds-ld.htm

EPA (Environmental Protection Agency). 1998d. Tier 2 Report to Congress. EPA 420/R-98–008. Office of Air and Radiation, U.S. Environmental Protection Agency, Washington, DC.

EPA (U.S. Environmental Protection Agency). 1998e. Air Conditioning Correction Factors in MOBILE6. Draft. M6.ACE.002. Office of Mobile Sources. U.S. Environmental Protection Agency. Available: http://www.epa.gov/omswww/models.htm

EPA (U.S. Environmental Protection Agency). 1998f. Air Conditioning Activity Effects in MOBILE6. Draft. M6.ACE.001. Office of Mobile Sources. U.S. Environmental Protection Agency. Available: http://www.epa.gov/omswww/models.htm.

EPA (U.S. Environmental Protection Agency). 1998g. Hot Soak Emissions as a Function of Soak Time. EPA420-P-98–018. M6.EVP.007. Office of Mobile Sources. U.S. Environmental Protection Agency.

EPA (U.S. Environmental Protection Agency). 1998h. Fuel Oxygen Effects on Exhaust CO Emissions. Recommendations for MOBILE6. Draft. EPA420-P-98–006. M6. FUL.002. Office of Mobile Sources. Office of Air and Radiation. U.S. Environmental Protection Agency.

EPA (U.S. Environmental Protection Agency). 1998i. Update Heavy-Duty Engine Emission Conversion Factors for MOBILE6: Analysis of BSFCs and Calculation of Heavy-Duty Engine Emission Conversion Factors. EPA420-P-98–015. M6.HDE.004. Office of Mobile Sources, Office of Air and Radiation. U.S. Environmental Protection Agency. May.

EPA (U.S. Environmental Protection Agency). 1998j. AP-42, Air Pollutant Emission Factors, 1998. Office of Mobile Sources. U.S. Environmental Protection Agency. Available: http://www.epa.gov/ttn/chief/ap42.html

EPA (U.S. Environmental Protection Agency). 1998k. Methodology for Modifying MOBILE5b in the Tier 2 Study. EPA-R-98–004. U.S. Environmental Protection Agency. April.

EPA (U.S. Environmental Protection Agency). 1999a. Control of Air Pollution from New Motor Vehicle Standards: Proposed Tier 2 Motor Vehicle Emissions Standards and Gasoline Sulfur Control Requirements. Final Rule, Fed. Regist. 64(92):26003.

EPA (U.S. Environmental Protection Agency). 1999b. Draft Guidance on the Use of Models and Other Analyses in Attainment Demonstrations for the 8-Hour Ozone NAAQS, EPA-454/R-99–004. Office of Air Quality Planning and Standards, U.S. Environmental Protection Agency, Research Triangle Park, NC.

EPA (U.S. Environmental Protection Agency). 1999c. Development of Light-Duty Emissions Inventory Estimates in the Notice of Proposed

Rulemaking for Tier 2 and Sulfur Standards. EPA420-R-99–005. Office of Mobile Sources, Office of Air and Radiation. U.S. Environmental Protection Agency.

EPA (U.S. Environmental Protection Agency). 1999d. Regulatory Impact Analysis: Control of Air Pollution from New Motor Vehicles: Tier 2 Motor Vehicle Emissions Standards and Gasoline Sulfur Control Requirements. EPA-420/D-99–001. Office of Air and Radiation and Office of Mobile Sources. U.S. Environmental Protection Agency, Ann Arbor, MI.

EPA (U.S. Environmental Protection Agency). 1999e. Description of the MOBILE Highway Vehicle Emissions Factor Model. Office of Mobile Sources, Anne Arbor, MI. Online. Available: http://www.epa.gov/oms/ models/mdlsmry.txt

EPA (U.S. Environmental Protection Agency). 1999f. Estimation of Motor Vehicle Toxic Emissions and Exposure in Selected Urban Areas. Vol. I. Draft. EPA420-D-99–002a. Office of Mobile Sources, Office of Air and Radiation. U.S. Environmental Protection Agency. March.

EPA (U.S. Environmental Protection Agency). 1999g. Determination of Running Emissions as a Function of Mileage for 1981–1993 Model Year Light-Duty Cars and Trucks. Draft. EPA 420-P-99–010. M6.EXH.001. Office of Mobile Sources. Office of Air and Radiation. U.S. Environmental Protection Agency. March.

EPA (U.S. Environmental Protection Agency). 1999h. Determination of Start Emissions as a Function of Mileage and Soak Time for 1981–1993 Model Year Light-Duty Vehicles. Draft. EPA420-P-99–015. M6.STE. 003. Office of Mobile Sources, Office of Air and Radiation. U.S. Environmental Protection Agency. March.

EPA (U.S. Environmental Protection Agency). 1999i. Estimating Weighting Factors for Evaporative Emissions in MOBILE6 Draft. EPA420-P-99–023. M6.EVP.006. Office of Mobile Sources, Office of Air and Radiation. U.S. Environmental Protection Agency. June.

EPA (U.S. Environmental Protection Agency). 1999j. Facility-Specific Speed Correction Factors. Draft. EPA420-P-99–002. M6.SPD.002. Office of Mobile Sources, Office of Air and Radiation. U.S. Environmental Protection Agency. August.

EPA (U.S. Environmental Protection Agency). 1999k. The Determination of Hot Running Emissions from FTP Bag Emissions. EPA420-P-99–014. M6.STE.002. Office of Mobile Sources. U.S. Environmental Protection Agency. Available: http://www.epa.gov/oms/ m6.htm

EPA (U.S. Environmental Protection Agency). 1999l. Overview of Methodology for Tier 0 In-Use Deterioration and Key Issues for Comment. Draft. EPA420-P-99–016. M6.EXH.008. Office of Mobile Sources, Office of Air and Radiation. U.S. Environmental Protection Agency. March.

EPA (U.S. Environmental Protection Agency). 1999m. Analysis of Emis sions Deterioration Using Ohio and Wisconsin IM240 Data. Draft. EPA420-P-99–013. M6.EXH.002. Office of Mobile Sources, Office of Air and Radiation. U.S. Environmental Protection Agency. March.

EPA (U.S. Environmental Protection Agency). 1999n. MOBILE6 Inspection/Maintenance Benefits Methodology for 1981 through 1993 Model Year Light Vehicles. Draft. EPA420-P-99–007. M6.IM.001. Office of Mobile Sources, Office of Air and Radiation. U.S. Environmental Protection Agency.

EPA (U.S. Environmental Protection Agency). 1999o. Determination of CO Basic Emission Rates, OBD and I/M Effects for Tier I and Later LDVs and LDTs. Draft. EPA420-P-99–017. M6.EXH.009. Office of Mobile Sources, Office of Air and Radiation. U.S. Environmental Protection Agency.

EPA (U.S. Environmental Protection Agency). 1999p. Determination of NO_X and HC Basic Emission Rates, OBD and I/M Effects for Tier I and Later LDVs and LDTs. Draft. EPA420-P-99–009. M6.EXH.007. Office of Mobile Sources, Office of Air and Radiation. U.S. Environmental Protection Agency. March.

EPA (U.S. Environmental Protection Agency). 1999q. Evaluating Resting Loss and Diurnal Evaporative Emissions Using Real Time Diurnal Tests. Draft. EPA420-P-99–026. M6EVP.001. Office of Mobile Sources, Office of Air and Radiation. U.S. Environmental Protection Agency.

EPA (U.S. Environmental Protection Agency). 1999r. Estimating Running Loss Evaporative Emissions in MOBILE6. Draft. EPA420-P-99– 024. M6.EVP.008. Office of Mobile Sources, Office of Air and Radiation. U.S. Environmental Protection Agency. June.

EPA (U.S. Environmental Protection Agency). 1999s. Fuel Sulfur Effects on Exhaust Emissions: Recommendations for MOBILE6. EPA420-P-99– 008. M6.FUL.001. Office of Mobile Sources, Office of Air and Radiation. U.S. Environmental Protection Agency. March.

EPA (U.S. Environmental Protection Agency). 1999t. Air Pollution; Tier 2 Motor Vehicle Emission Standards and Gasoline Sulphur Control Requirements. Fed. Regist. 64(92):26003–26142. May, 13.

EPA (U.S. Environmental Protection Agency). 1999u. Exhaust Emission Temperature Correction Factors for MOBILE: Engine Start and Running LA4 Emissions for gasoline Vehicles. Draft. EPA420-P-99–001. M6. STE.004. Office of Mobile Sources, Office of Air and Radiation. U.S. Environmental Protection Agency. January.

EPA (U.S. Environmental Protection Agency). 1999v. Update of Heavy-Duty Emission Levels (Model Years 1988–2004+) for Use in MOBILE6. EPA420-R-99–010. M6.HDE.001. Office of Mobile Sources, Office of Air and Radiation. U.S. Environmental Protection Agency. April.

EPA (U.S. Environmental Protection Agency). 1999w. Development and Use of Heavy-Duty Defeat Device Emission Effects for MOBILE5 AND

MOBILE6. EPA420-P-99–030. M6.HDE.003. Office of Mobile Sources, Office of Air and Radiation. U.S. Environmental Protection Agency. October.

EPA (U.S. Environmental Protection Agency). 1999x. Fleet Characterization Data for MOBILE6: Development and Use of Age Distributions, Average Annual Mileage Accumulation Rates and Projected Vehicle Counts for Use in MOBILE6. EPA420-P-99–011. M6.FLT.007. Office of Mobile Sources, Office of Air and Radiation. U.S. Environmental Protection Agency. April.

FHWA (Federal Highways Administration). 1992. A Transportation Programs and Provisions of the Clean Air Act Amendments of 1990. FHWA-PD-92–023. Washington DC: U.S. Department of Transportation.

FHWA (Federal Highway Administration). 1998. Traffic Software Integrated System Version 4.2 User's Guide, ITT Systems and Sciences Corporation. Available: http://www.fhwa-tsis.com/

Fox, J.W. 1996. Air Pollution Abatement: Quantifying Automobile Emissions. Ph.D. Thesis Department of Mechanical Engineering. Massachusetts Institute of Technology. September.

FRG-FEPA (Federal Republic of Germany-Federal Ministry for the Environment). 1993. Exhaust Emission Factors for Motor Vehicle in the Federal Republic of Germany for the Reference Year 1990. UFO Plan No. 140 05 152 and 104 05 509, UFO-FO 91–042. Environmental Research Plan for the Federal Ministry for the Environment, Nature Conservation and Reactor Safety by Institute of Environmental Protection and Energy Technology.

Fujita, E. 1999. Application of Field Measurements to Evaluate Emission Inventory Estimates. Presented at the workshop of the Committee to Review EPA's Mobile-Source Emissions Factor (MOBILE) Model, March 4, 1999. Irvine, CA.

Fujita, E.M., Z.Lu, J.C.Sagebiel, N.F.Robinson, and J.G.Watson. 1995. VOC Source Apportionment for the Coastal Oxidant Assessment for Southeast Texas. Desert Research Institute, Reno, NV. Prepared for the Texas Natural Resource Conservation Commission, Austin, TX.

Fujita, E.M., J.G.Watson, J.C.Chow, N.Robinson, L.Richards, and N. Kumar. 1998. Northern Front Range Air Quality Study. Volume C: Source Apportionment and Simulation Methods and Evaluation. Final report. Colorado State University, Fort Collins, CO. Available: http://nfraqs.cira.colostate.edu/finals.html

GAO (General Accounting Office). 1996. Peer Review B EPA's Implementation Remains Uneven: Report to Congressional Requesters. Washington, D.C.: U.S. General Accounting Office

GAO (General Accounting Office). 1997. Air Pollution: Limitations of EPA's Motor Vehicle Emissions Model and Plans to Address Them:

Report to the Chairman, Subcommittee on Oversight and Investigations, House of Representatives. Washington, D.C.: U.S. General Accounting Office.

Gertler, A.W. J.C.Sagebiel, D.N.Wittorff, W.R.Pierson, W.A.Dippel, D. Freeman, and L.Sheetz. 1997. Vehicle Emissions in Five Urban Tunnels. NTIS/PB97–158091. Nevada Univ. System, Reno: Desert Research Inst.

Gorse, R.A. 1999. In-use Phoenix vehicle evaporative emissions. Pp. 2.15– 2.21 in: Proceedings of the 9th On-Road Vehicle Emissions Workshop, April 19–21, 1999. San Diego, CA.

Graboski, M.S., R.L.McCormick, J.Yanowitz and L.Ryan. 1998. Heavy-Duty Vehicle Testing for the Northern Front Range Air Quality Study. Final Report. Colorado Institute for Fuels and High-Altitude Engine Research, Colorado School of Mines, Golden, CO. Prepared for The Office of the Vice President for Research and Information Technology, Colorado State University, Fort Collins, CO. Feb. 24.

Guensler, R. 1993. Vehicle Emission Rate and Average Vehicle Operating Speeds. Ph. D. dissertation. University of California, Davis, CA.

Guensler, R., and J.D.Leonard II. 1997. MOBILE5a Confidence Intervals for Average Speed Emission Rates. School of Civil and Environmental Engineering, Georgia Institute of Technology. Atlanta, GA.

Guensler, R., S.Washington and W.Bachman. 1998. Overview of the MEASURE Modeling Framework. Pp. 51–70 in Transportation Planning and Air Quality III: Emerging Strategies and Working Solution. Presented at Transportation Research Board 77th Annual Meeting. Jan.

Harley, R.A., A.G.Russell, and G.R.Cass. 1993a. Mathematical modeling of the concentrations of volatile organic compounds: Model performance using a lumped chemical mechanism. Environ. Sci. Technol. 27:1638– 1649.

Harley, R.A., A.G.Russell, G.J.McRae, G.R.Cass, and J.H.Seinfeld. 1993b. Photochemical modeling of the Southern California Air Quality Study. Environ. Sci. Technol. 27:378–388.

Harrington, W., V.D.McConnell and M.Cannon. 1998. Behavior Analysis of EPA's MOBILE Emissions Factor Model. DP-98–7. Washington, D.C.: Resources for the Future.

Harvey, G. and E.Deakin. 1993. A Manual of Regional Transportation Modeling Practice for Air Quality Analysis, Version 1.0. National Association of Regional Councils.

Haskew, H.M., E.King, T.F.Liberty and R.R.Reuter. 1999. Running Loss Emissions from In-Use Vehicle. SAE Technical Paper 1999–01– 1464. Society of Automotive Engineers, Warrendale, PA.

Heck, W.W., O.C.Taylor, R.Adams, G.Bingham, J.Miller, E.Preston, and L.Weinstein. 1982. Assessment of crop loss from ozone. J. Air Pollut. Control Assoc. 32:353–361.

Heiken, J., J.L.Fieber, S.B.Shepard, J.P.Cohen, A.K.Pollack, and G.Z. Whitten. 1994. Investigation of MOBILE5a Emission Factors: Assessment of Exhaust and Nonexhaust Emission Factor Methodologies and Oxygenate Effects. Health and Environmental Sciences Department. API Pub. No. 4603. Prepared by Systems Applications International, San Rafael, CA. For American Petroleum Institute. February.

Heinz Center. 1998. Designs for Domestic Carbon Emissions Trading. The H.John Heinz III Center for Science, Economics and the Environment, Washington, D.C. Available: http://www.heinzctr.org/publications/index.htm.

Holmes, K.J., and H.Ellis. 1997. Simulation of halocarbon production and emissions and effects on ozone depletion. Environ. Manage. 21(5): 669–685.

Hyde, J.D. 1998. Influence of Gasoline RVP on Carbon Monoxide Emissions New York Department of Environmental Conservation, Division of Air Resources, Bureau of Mobile Sources, Albany, N.Y. Dec. 15.

IFEU (Institut für Energie- und Umweltforschung, Heidelberg.) 1997. Daten- und Rechenmodell: Energieverbrauch und Schadstoffemissionen des motorisierten Verkehrs in Deutschland 1980–2020. On behalf of the German Federal Environmental Agency (UBA), UFOPLAN 105–06–057, Berlin.

Ingalls, M.N., L.R.Smith, and R.E.Kirksey. 1989. Measurement of On-Road Vehicle Emission Factors in the California South Coast Air Basin. Vol. 1: Regulated Emissions. PB898220925XSP. Southwest Research Inst., San Antonio, TX. Prepared for the Coordinating Research Council, Atlanta, GA. June.

Jimènez -Palacios, J.L. 1999. Understanding and Quantifying Motor Vehicle Emissions with Vehicle Specific Power and TILDAS Remote-Sensing. Ph. D. dissertation. Massachusetts Institute of Technology, Cambridge, MA.

Jimènez, J., P.PcClintock, J.G.Mcrae, D.D.Nelson and M.S.Zahniser. 1999. Vehicle specific power: A useful parameter for remote sensing and emission studies. Pp. 7.43–7.57 in: Proceedings of the 9th On-Road Vehicle Emissions Workshop, April 19–21. San Diego, CA.

Joumard, R. P.Jost, J.Hickman and D.Hassel. 1995. Hot passenger car emissions modelling as a function of instantaneous speed and acceleration. Sci. Total Environ. 169(1–3):167–174.

Keenan, M. and D.Escarpeta. 1994. Development of a base year on-road mobile source emission inventory for New York State and its applications for UAM modeling and examining VOC/NOX emission ratios. Pp. 105–124 in: The Emission Inventory: Applications and Improvement, Proceedings of an International Specialty Conference, Raleigh, NC, November 1–3, 1994. Pittsburgh, PA: Air & Waste Management Association.

Kini, M.D. and H.C.Frey. 1997. Probabilistic Evaluation of Mobile Source Air Pollution. Vol. I. Probabilistic Modeling of Exhaust Emissions from Light Duty Gasoline Vehicles. Final Report. PB99–146292. The Center for Transportation and the Environment, Raleigh, NC. Dec.

Knepper, J.C., W.J.Koehl, J.D.Benson, V.R.Burns, R.A.Gorse Jr., A.M. Hochhauser, W.R.Leppard, L.A.Rapp, and R.M.Reuter. 1993. Fuel Effects in Auto/Oil High Emitting Vehicles. SAE Technical Paper 930137. Society of Automotive Engineers, Warrendale, PA.

Korc, M., C.M.Jones, L.R.Chinkin, H.H.Main, P.T.Roberts and C.Blanchard. 1995. Use of PAMS Data to Evaluate the Texas Coast Emission Inventory. Final Report. Prepared for EPA, Research Triangle Park, NC. Work Assignment 2–95, EPA Contract 68D30020, STI-94520–1558-FR. Sonoma Technology, Inc., Santa Rosa, CA.

Lakshminarayanan, A. 1999. Analysis of the Sensitivity of Photochemical Airshed Modeling to Grid Size and Spatial and Temporal Distributions of Mobile Source Emissions, Masters Thesis, Civil and Environmental Engineering, Georgia Institute of Technology, Atlanta, GA. July.

LANL (Los Alamos National Laboratory). 1999. Transportation Analysis Simulation System (TRANSIMS), Vol.0-Overview, LANL Tech. report LA-UR-99–1658, Los Alamos National Laboratory. University of California. April.

Lawson, D.R., P.J.Groblicki, D.H.Stedman, G.A.Bishop, and P.L. Guenther. 1990. Emissions from in-use motor vehicles in Los Angeles: A pilot study of remote sensing and the inspection and maintenance program. J. Air Waste Manage. Assoc. 40(8):1096–1105.

Lax, D. 1994. The Effects of Fuel RVP and Fuel Blends on Emissions at Non-FTP (Federal Test Procedure) Temperatures, Vol. 1: Summary Report. Health and Environmental Sciences Dept. Report 4533. Washington D.C.: American Petroleum Institute. August.

Lippmann, M. 1989. Health effects of ozone: A critical review. J. Air Waste Manage. Assoc. 39:672–695.

McClement, D. 1999. Real World Evaporative Testing of Late Model In-Use Vehicles, CRC Project E-41, Prepared by Automotive Testing Laboratories Inc., Mesa, AZ. for the Coordinating Research Council, Atlanta, GA. Oct. 4.

McClement, D., J.A.Dueck and B.Hall. 1998. Measurement of Running Loss Emissions from In-Use Vehicles. CRC Project E-35. CRC Rep. No. 611, PB99–158024INZ. Prepared by Automotive Testing Labs, Inc., Mesa, AZ. for the Coordinating Research Council, Atlanta, GA.

McClintock, P. 1999. I/M Impact on Dayton Registrations. Memo to Phil Lorang, EPA OMS. June, 25.

Milkins, E. and H.Watson. 1983. Comparison of Urban Driving Patterns. SAE Technical Paper No. 830939. Society of Automotive Engineers, Warrendale, PA.

NAPAP (National Acid Precipitation Assessment Program). 1990. Effects of acidic deposition on materials. NAPAP Report 19. Pp. 3–280 in: Acid Deposition: State of Science and Technology, Vol. III. Terrestrial, Materials, Health and Visibility Effects, P.M.Irving, ed. Washington, D.C.: NAPAP.

NCTCOG (North Central Texas Council of Governments). 1998. Air Quality Conformity Analysis. NCTCOG, Arlington, TX.

NCTCOG (North Central Texas Council of Governments). 1999. Dallas-Fort Worth Regional Travel Model: Description of the Mulitmodal Forecasting Process. NCTCOG, Arlington, TX.

Niemeier, D.A., J.M.Utts, K.Nanzetta, and L.Ring. 1998. Speed Correction Curves: Scoping Study. Report UCD-ITS-RR-98–12. Institute of Transportation Studies, University of California, Davis, CA. October.

Norbeck, J.M., T.D.Durbin and T.J.Truex. 1998. Measurement of Primary Particulate Matter Emissions from Light-Duty Motor Vehicles. CRC Project No. E-24–2. Final Report. Center for Environmental Research and Technology, University of California, Riverside, CA for Coordinating Research Council and South Coast Quality Management District. December. Available: http://www.crcao.com/new.htm

Norton, T., S.Tucker, R.E.Smith and D.R.Lawson. 1998. The Northern Front Range Air Quality Study. EM (January):13–19.

NRC (National Research Council). 1991. Rethinking the Ozone Problem in Urban and Regional Air Pollution. Washington, D.C.: National Academy Press.

NRC (National Research Council). 1993. Protecting Visibility in National Parks and Wilderness Areas. Washington, D.C.: National Academy Press.

NRC (National Research Council). 1996. Toxicological and Performance Aspects of Oxygenated Motor Vehicle Fuels. Washington, D.C.: National Academy Press.

NRC (National Research Council). 1999. Ozone-Forming Potential of Reformulated Gasoline. Washington, D.C.: National Academy Press.

NSTC (National Science and Technology Council). 1997. Interagency Assessment of Oxygenated Fuels. National Science and Technology Council, Committee on Environment and Natural Resources, Office of Science and Technology Policy, Executive Office of the President of the United States.

Ott, W.R. 1985. Total human exposure. Environ. Sci. Technol. 19:880–886.

Paramics. 1998. The Paramics Transportation Modeling Suite. Available: html://www.paramics-online.com/.

Pedersen, N.J. and D.R.Samdahl. 1982. Highway Traffic Data for Urbanized Area Project Planning and Design, NCHRP Report 255, Transportation Research Board, National Research Council, Washington, D.C.

Pemberton, M. 1996. World Vehicle Forecasts and Strategies: The Next

20 Years: A Special Report Covering the Period from 1960–2015. Southfield, MI: Ward's Communications.

Pierson, W.R, A.W.Gertler, and R.L.Bradow. 1990. Comparison of the SCAQS Tunnel study with other on-road vehicle emission data. J. Air Waste Manage. Assoc. 40(11):1495–1504.

Pollack, A.K. and C.E.Lindhjem. 1998. Nonroad mobile emissions modeling. Pp. 640–650 in Emission Inventory: Planning for the Future, Proceedings of a Specialty Conference, Research Triangle Park, NC. Oct. 28–30, 1997. Pittsburgh, PA: Air & Waste Management Association.

Pollack, A., T.Stoeckenius and C.Tran. 1999a. MOBILE6 Deterioration Rates Memorandum to Phil Enns, EPA OMS. August, 18.

Pollack, A.K., P.Bhave, J.Heiken, K.Lee, S.Shepard, C.Tran, G.Yarwood, R.F.Sawyer, and B.A.Joy. 1999b. Investigation of Emission Factors in the California EMFAC7G Model. Final Report for CRC Project E-39, PB99–149718INZ. Prepared by ENVIRON International Corp., Novato, CA for Coordinating Research Council, Atlanta, GA.

Pollack, A.K., J.L.Fieber, J.Heiken, B.Austin, L.Chinkin, and M.Smylie. 1991. Assessment of Computer Models for Estimating Vehicle Emission Factors, Vol. I: Executive Summary, Vol. II: Description and Comparison of the MOBILE4 and EMFAC7E Models, Vol. III: Sensitivity Analyses of the MOBILE4 and EMFAC7E Emission Models. Prepared for Coordinating Research Council B APRAC., Atlanta, GA.

Ragazzi, R. and K.Nelson. 1999. The Impact of a 10% Ethanol Blended Fuel on the Exhaust Emissions of Tier 0 and Tier 1 Light Duty Gasoline Vehicles at 35°F, Colorado Department of Public Health and Environment, Denver, CO.

Reuter, R.M., R.A.Gorse, Jr., L.J.Painter, J.D.Benson, A.M.Hochhauser, B.H.Rippon, V.R.Burns, W.J.Koehl, and J.A.Rutherford. 1992. Effects of oxygenated fuels and RVP on automotive emissions. Auto/Oil Air Quality Improvement Research Program. SAE Technical Paper No. 920326. Pp. 391–412 in: Auto/Oil Air Quality Improvement Research Program, SAE SP-920. Society of Automotive Engineers, Warrendale, PA. February.

Scherrer, H.C. and D.B.Kittelson. 1994. I/M Effectiveness as Directly Measured by Ambient CO Data, SAE Technical Paper 940302. Society of Automotive Engineers, Warrendale, PA.

Sierra Research, Inc. 1994a. Investigation of MOBILE5a Emission Factors: Evaluation of I/M and LEV Program Benefits. Health and Environmental Sciences API Pub. No. 4604. Washington, D.C.: American Petroleum Institute. June.

Sierra Research, Inc. 1994b. Investigation of MOBILE5a Emission Factors: Evaluation of IM240-to-FTP Correlation and Base Emission Rate Equations. Health and Environmental Sciences API Pub. No. 4605. Washington, D.C.: American Petroleum Institute. June.

Sierra Research, Inc. 1994c. Evaluation of MOBILE Vehicle Emission Model. Prepared for J.A.Volpe National Transportation Systems Center and the U.S. Department of Transportation. December.

Singer, B.C. and R.A.Harley. 1996. A fuel-based motor vehicle emission inventory. J. Air Waste Manage. Assoc. 46(6):581–593.

Singer, B.C., and R.A.Harley. 2000. A fuel-based inventory of motor vehicle exhaust emissions in the Los Angeles area during summer 1997. Atmos. Environ. 34(11):1783–1795.

Singer, B., T.W.Kirchstetter, R.A.Harley, G.R.Kendall and J.M.Hesson. 1999. A fuel-based approach to estimating motor vehicle cold start emissions. J. Air Waste Manage. Assoc. 49 (2):125–135.

Shih, C. 1990. MOBILE4 Sensitivity Analysis. Technical Report. EPA-AA-TEB-90–02. U.S. Office of Mobile Sources. Office of Air and Radiation. U.S. Environmental Protection Agency. May.

Stedman, D.H. 1989. Automobile carbon monoxide emission. Environ. Sci. Technol. 23(2) 147–148.

Stedman, D.H., G.A.Bishop, P.Aldrete, and R.S.Slott. 1997. On-road evaluation of an automobile emission test program. Environ. Sci. Technol. 31:927–931.

Stedman, D.H., G.A.Bishop, R.S.Slott. 1998. Repair avoidance and evaluating inspection and maintenance programs. Environ. Sci. Technol. 32(10):1544–1545.

Sturm, P.J., G.Kirchweger and R.A.Almbauer. 1998. Instantaneous emission data and their use in estimating road traffic emissions. International Journal of Vehicle Design 20(1/4):181–191.

Sturm, P.J., R.A.Almbauer, C.Sudy and K.Pucher. 1997. Application of computational methods for the determination of traffic emissions. J. Air Waste Manage. Assoc. 47(11):1204–1210.

UBA/SAEFL (German Federal Environmental Agency/Swiss Agency for the Environment, Forests and Landscape). 1999. HBEFA-Handbook Emission Factors for Road Transport, Version 1.2. Berlin/Bern: UBA/ SAEFL. Jan.

U.S. Congress. 1995. Hearings before the Subcommittee on Oversight and Investigations of the Committee on Commerce, House of Representatives: 104th Congress, First Session on Inspection and Maintenance Programs. Clean Air Act Amendments, Serial No. 104–16. March 23–24. Washington, DC.: U.S. Government Printing Office.

Vallette, J. 1995. Deadly Complacency: US CFC Production, the Black Market, and Ozone Depletion. Washington, D.C.: Ozone Action.

Van Aerde, M. & Associates. 1995. "INTEGRATION Release 2.0 User's Guide," Kingston, Ontario, Canada.

Wagner, K.K., and N.J.M.Wheeler. 1993. Multi-species evaluation of the Urban Airshed Model with the SCAQS database. Pp. 264–269. In: Southern California Air Quality Study. Data Analysis. Proceedings of the International Specialty Conference. Pittsburgh, PA: Air & Waste Management Association.

Washington, S., J.Wolf and R.Guensler. 1997. Binary recursive partitioning method for modeling hot-stabilized emissions from motor vehicles. Transportation Research Record 1587:96–105.

Watson, J.G., E.Fujita, J.C.Chow, B.Zielinska, L.W.Richards, W.Neff, and D.Dietrich. 1998. Northern Front Range Air Quality Study. Final report. DRI 6580–685–8750.1F2. Prepared for The Office of the Vice President for Research and Information Technology, Colorado State University, Fort Collins, CO. June, 30. Available: http://nfraqs.cira. colostate.edu/dri.html

Watson, J.G., J.C.Chow, and E.M Fujita. In press. Review of volatile organic compound source appointment by chemical mass balance. Atmospheric Environment. In press

Weaver, C.S., M.A.Balam, C.J.Brodrick, C.Tran, J.Heiken and A.Pollack. 1998. Modeling Deterioration in Heavy-Duty Diesel Particulate Emissions. Final Report. Prepared for Office of Mobile Sources, US EPA. Engine, Fuel and Emissions Engineering, Incorporated, Sacramento, CA. June.

Wenzel, T. 1997. In-use Emissions and Failure Rates by Vehicle Model. Proceedings of the 7th CRC On-Road Vehicle Emissions Workshop, San Diego, CA. April.

Wenzel, T. 1999. Evaluation of Arizona's Enhanced I/M Program. Presentation to NRC Committee to Review MOBILE Model, March 4, 1999.

Wenzel, T., B.Singer, and R.Slott. In Press. Statistical Challenges of Analyzing Vehicle Emissions. Journal of Transporation and Statistics. In press.

West, B. and R McGill. 1997. Emissions and Fuel Consumption Predictions from Data-Based Modal Models. Proceedings of the 7th CRC On-Road Vehicle Emissions Workshop, San Diego, CA. April.

Williams, R.L. and S.H.Cadle. 1978. Characterization of tire emissions using an indoor test facility. Rubber, Chemistry and Technology. 51(1):7–25.

Williams, M.D., G.Thayer and M.J.Barthand L.L.Smith. 1999. The TRANSIMS Approach to Emission Estimation. LA-UR 99–471. New Mexico: Los Alamos National Laboratory. Available: http://transims. tsasa.lanl.gov/

Wolff, G.T. 1996. The scientific basis for a new ozone standard. EM (September):27–32.

Zhang, Y., G.A.Bishop, and D.H.Stedman. 1994. Automobile emissions are statistically gamma-distributed. Environ. Sci. Tech. 28(7):1370– 1374.

Zielinska, B., J.Sagebiel, G.Harshfield, and E.Fujita. 1999.. Air Monitoring Program for Determination of the Impacts of the Introduction of

California's Phase 2 Reformulated Gasoline on Ambient Air Quality in the South Coast Air Basin. Energy and Environmental Engineering Center, Desert Research Inst. Reno, NV. Prepared for the California State Air Resources Board, Sacramento, CA. February.

Glossary

Air-quality model— A computer-based mathematical model used to predict air quality based upon emissions and the effects of the transport, dispersion, and transformation of compounds emitted into the air.

Arterial— A roadway that serves major traffic movements, and secondarily provides access to abutting land. Arterials generally carry higher traffic volumes at higher speeds than collectors and local streets, but carry lower volumes at lower speeds than expressways, freeways, and other limited access and grade separated facilities.

Ambient air— The air outside of structures. Often used interchangeably with "outdoor air."

California Air Resources Board (CARB)— A part of the California Environmental Protection Agency whose mission it is to promote and protect public health, welfare and ecological resources through the effective and efficient reduction of air pollutants while recognizing and considering the effects on the economy of the state.

Carbon monoxide (CO)— A colorless, odorless gas resulting from the incomplete combustion of hydrocarbon fuels.

Clean Air Act (CAA)— The original Clean Air Act was passed in 1963, but our national air pollution control program is actually based on the 1970 version of the law. The 1990 Clean Air Act Amendments (CAAA90) are the most recent and far-reaching revisions of the 1970 law.

Clean screening— The use of methods such as remote sensing measurements or vehicle profiling by states to excuse cars from a scheduled inspection and maintenance (I/M) emissions test.

Collector— A street that provides access within neighborhoods as well as commercial and industrial districts, and which channels traffic from local streets to arterials.

Conformity (transportation conformity)— A process to demonstrate whether a federally supported activity is consistent with the air quality goals in State Implementation Plans (SIPs). Transportation conformity demonstrates that plans, programs, and projects approved or funded by the Federal Highway Administration, or the Federal Transit Administration for regionally-significant projects do not create new violations, increase the frequency or severity of existing violations, or delay timely attainment of NAAQS. General conformity refers to projects approved or funded by other federal agencies.

Criteria air pollutants—A group of common air pollutants (carbon monoxide, lead, nitrogen dioxide, ozone, particulate matter, and sulfur dioxide) regulated by the Federal Government since the passage of the Clean Air Act in 1970 on the basis of information on health and/or environmental effects of each pollutant.

Dose— The amount of a contaminant that is absorbed or deposited in the body of an exposed organism for an increment of time-usually from a single medium. Total dose is the sum of doses received by a person from a contaminant in a given interval resulting from interaction with all environmental media that contain the contaminant. Units of dose and total dose (mass) are often converted to units of mass per volume of physiological fluid or mass of tissue.

Dose-response— The relationship between the dose of a pollutant and the response or effect it produces on a biological system.

Emissions budget— Allowable emissions levels identified as part of a state implementation plan for pollutants emitted from mobile, industrial, stationary, and area sources. These emissions levels are used for meeting emission reduction milestones, attainment, or maintenance demonstra-tions.

Emissions factor— The relationship between the amount of pollution produced and the amount of raw material processed or burned, or between the amount of pollution produced and the activity level. By using the emission factor of a pollutant and specific data regarding quantities of materials used by a given source or the activity of a given source, it is possible to compute emissions for the source. In the case of mobile source emissions, the product of an emission factor estimated in mass of pollutant per unit distance (e.g., grams per mile) and an activity estimate in distance (e.g., average miles traveled) produces an estimate of emissions. In the case of stationary source emissions, the product of an emission factor in mass of pollutant per unit energy (e.g., pounds per million Btu) and the amount of energy consumed produces an estimate of emissions.

Emissions inventory— Estimates of the amount of pollutants emitted into the atmosphere from major mobile, stationary, area-wide, and natural source categories over a specific period of time such as a day or a year.

Exceedance— Air pollution event in which the ambient concentration of a pollutant exceeds a National Ambient Air Quality Standard (NAAQS).

Ethanol— Ethyl-alcohol, a volatile alcohol containing two carbon atoms (CH_3CH_2OH). For fuel use, ethanol is produced by fermentation of corn or other plant products.

Exposure— An event that occurs when there is contact at a boundary between a human and the environment with a contaminant of a specific concentration for an interval of time; the units of exposure are concentration multiplied by time.

Fast pass— Fast pass is a process that recognizes very clean cars early in the IM240 test cycle and passes them without the need to complete the full test.

Federal Implementation Plan (FIP)— In the absence of an approved state implementation plan, a plan prepared by the U.S. Environmental Protection Agency (EPA) that provides measures that nonattainment areas must take to meet the requirements of the Federal Clean Air Act.

Federal Test Procedure (FTP)— A certification test for measuring the tailpipe and evaporative emissions from new vehicles over the Urban Dynamometer Driving Schedule, which attempts to simulate an urban driving cycle.

Gasoline volatility— The evaporative properties of gasoline. Gasoline vapor contains volatile organic compounds.

Gross Vehicle Weight Rating (GVWR)— The value specified by the manufacturer as the maximum design loaded weight of a single vehicle (i.e., vehicle weight plus rated cargo capacity).

Heavy-duty vehicles (HDV)— Any motor vehicle rated at more than 8,500 pounds gross vehicle weight GVWR or that has a vehicle curb weight of more than 6,000 pounds or that has a basic vehicle frontal area in excess of 45 square feet. This excludes vehicles that will be classified as medium-duty passenger vehicles for the purposes of the Tier 2 emissions standards.

Heavy-duty diesel vehicles (HDDV) — An HDV using diesel fuel.

Inspection and maintenance (I/M) program— A periodic emissions testing and inspection program, usually done once a year or once every two years, to ensure that the catalytic or other emissions control devices on in-use vehicles are operating. The test can be an idle or dynamometer loaded mode emission test.

Light-duty vehicle (LDV)— A passenger car or passenger car derivative capable of seating 12 or fewer passengers. All vehicles and trucks under 8,500 GVWR are included (this limit previously was 6,000 pounds). Small pick-up trucks, vans, and sport utility vehicles may be included.

Medium-duty passenger vehicle (MDPV)— A new class of vehicles introduced with the Tier 2 emissions standards that includes sport utility vehicles and passenger vans rated at between 8,5000 and 10,000 GVWR.

Metropolitan Planning Organization (MPO) — The organized entity designated by law with lead responsibilities for developing transportation plans and programs for urbanized areas with population of 50,000 or more people. MPOs are established by agreement of the Governor and units of general purpose local government which together represent 75 percent of the affected population of an urbanized area.

Mode choice— A process by which an individual selects a transportation mode (e.g., automobile, transit, bicycle) for use on a trip, given the trip's purpose, origin, and destination.

National Low Emission Vehicle (NLEV) Vehicles that meet voluntary low emissions tailpipe standards that are more stringent than can be mandated by EPA prior to model year 2004. The NLEV program introduces California low emissions cars and light-duty trucks into the Northeast beginning in model year 1999 vehicles and the rest of the country in the model year 2001 vehicles.

National Ambient Air Quality Standards (NAAQS)— Standards set by EPA for the maximum levels of criteria air pollutants that can exist in the outdoor air without unacceptable effects on human health or the public welfare.

Nonattainment area— A geographic area in which the level of a criteria air pollutant is higher than the level allowed by the federal standards. A single geographic area may have acceptable levels of one criteria air pollutant but unacceptable levels of one or more other criteria air pollutants; thus, an area can be both attainment and nonattainment at the same time.

Nitrogen oxides (Oxides of Nitrogen, NO_x)— A general term pertaining to nitric oxide (NO), nitrogen dioxide (NO_2), and other oxides of nitrogen. Nitrogen oxides are typically created during combustion processes, and are major contributors to smog formation and acid deposition.

On-Board Diagnostic (OBD) systems— Devices that are incorporated into the computers of new motor vehicles to monitor fuel delivery and emission controls. The computer triggers a dashboard indicator light when the controls malfunction, alerting the driver to seek maintenance for the vehicle. Diagnostic systems are required on vehicles beginning with 1994 models.

Oxygenated gasoline (oxyfuel)— Gasoline containing oxygenates, typically methyl tertiary-butyl ether (MTBE) or ethanol, that burns more completely than regular gasoline and reduces production of carbon monoxide, a criteria air pollutant. In some parts of the country, carbon monoxide emissions from cars make a major contribution to pollution. In these areas, gasoline refiners must market oxygenated fuels, which contain a higher oxygen content than regular gasoline. For oxygenated gasoline programs to reduce carbon monoxide (CO) pollution, the minimum oxygen content is typically 2.7 weigh percent

Oxygenates— Compounds containing oxygen (alcohols and ethers) that are added to fuels to increase its oxygen content. Methyl tertiary-butyl ether (MTBE) and ethanol are the most common oxygenates currently used, although there are a number of other oxygenates.

Ozone— A reactive gas consisting of three oxygen atoms. It is a product of photochemical process involving the sunlight and ozone precursors, such as hydrocarbons and oxides of nitrogen. Ozone exists in the upper atmosphere ozone layer (stratospheric ozone) as well as at the earth's surface (tropospheric ozone). Tropospheric ozone causes plant damage and adverse health effects and is a criteria air pollutant. Tropospheric ozone is a major component of smog.

Particulate matter (PM)— Any material, except uncombined water, that exists in the solid or liquid state in the atmosphere. The size of participate matter can vary from coarse, wind-blown dust particles to fine particles directly emitted as combustion products or formed through secondary reactions in the atmosphere.

Photochemical reaction— A term referring to chemical reactions brought about by the light of the sun. The reaction of nitrogen oxides with hydrocarbons in the presence of sunlight to form ozone is an example of a photochemical reaction.

PM-2.5— A subset of particulate matter that includes those tiny particles with an aerodynamic diameter less than or equal to a nominal 2.5 micrometers. This fraction of particulate matter penetrates most deeply into the lungs, and causes the majority of visibility reduction.

PM-10— A major air pollutant consisting of small particles with an aerodynamic diameter less than or equal to a nominal 10 micrometers (about 1/7 the diameter of a single human hair). Their small size allows them to make their way to the air sacs deep within the lungs where they may be deposited and result in adverse health effects. PM10 also causes visibility reduction.

Preconditioning— A term used in inspection and maintenance programs that use the IM240 test cycle. The cut points, which determine passing or failing for such a vehicle, are based on testing a fully warmed-up vehicle in which the emissions control equipment, including the catalytic converter, are hot and fully functional. If an owner drives a short distance to the test station or if the vehicle has to wait in the test station for a long time, the vehicle may not be fully warmed up. This may result in a false reading; a car which would have passed if fully warmed (i.e., fully preconditioned) would fail. Thus, a preconditioned vehicle is a vehicle that is fully warmed up so that it can give a valid result from an IM240 inspection test.

Primary standard— A NAAQS for criteria air pollutants based on health effects.

Reformulated gasoline (RFG)— Specifically formulated fuels blended such that, on average, the exhaust and evaporative emissions of VOCs and hazardous air pollutants (chiefly benzene, 1,3-butadiene, polycyclic organic matter, formaldehyde, and acetaldehyde) resulting from RFG use in motor vehicles are significantly and consistently lower than such emissions resulting from use of conventional gasolines. The 1990 Clean Air Act Amendments requires sale of reformulated gasoline in the nine areas with the most severe ozone pollution problems. RFG contains, on average, a minimum of 2.0 weight percent oxygen.

Remote sensing— A method for measuring pollution levels in a vehicle's exhaust while the vehicle is traveling down the road. Remote sensing systems employ an infrared absorption principle to measure VOC and carbon monoxide emissions relative to carbon dioxide emissions. These systems typically operate by continuously projecting a beam of infrared radiation across a roadway. As a vehicle passes through the beam, the device measures the ratios of CO and VOC to carbon dioxide in the vehicle exhaust plume.

Reid vapor pressure (RVP)— fuel vapor pressure (often expressed in units of pounds per square inch) at 100 degrees F.

Secondary particle— Particulate matter that is formed in the atmosphere. Secondary particles are generally composed of species such as ammonia and the products of atmospheric chemical reactions including nitrates, sulfates and organic material. Secondary particles are distinguished from primary particles, which are emitted directly into the atmosphere.

Secondary standard— A NAAQS for criteria air pollutants based on environmental effects such as damage to property, plants, visibility, etc.

Speed-correction factor (SCF)— Factors used in the MOBILE model to adjust emissions factors from the average speed used in the Federal Test Procedure (which is used to obtain emissions data) to other average speeds as driven by vehicles in the geographical area being modeled.

Supplemental Federal Test Procedure (SFTP)— The SFTP is a certification test for measuring the tailpipe and evaporative emissions from new vehicles that includes two driving cycles not represented in the FTP. The SFTP includes a test cycle simulating high speed and high acceleration driving (US06 cycle) and a test cycle simulating air conditioner operation (SC03 cycle).

State implementation plan (SIP)—A detailed description of the programs a state will use to carry out its responsibilities under the Clean Air Act for complying with the NAAQS. SIPs are a collection of the programs used by a state to reduce air pollution. The Clean Air Act requires that EPA approve each SIP. The public is given opportunities to participate in review and approval of SIPs.

Two-way catalytic converter—First generation catalytic converters designed to reduce CO and VOC emissions from gasoline-fueled vehicles.

Three-way catalytic converter—Catalytic converters designed to reduce CO, VOC and NO_x emissions from gasoline-fueled vehicles

Tier 0 vehicles—Vehicles that meet Tier 0 tailpipe standards. For light-duty vehicles, these tailpipe standards began with model year 1981 vehicles and were phased out in model year 1995 for passenger cars and most light-duty trucks.

Tier 1 vehicles—Vehicles that meet Tier 1 tailpipe standards. For light-duty vehicles, these tailpipe standards began with model year 1994 vehicles.

Tier 2 vehicles—Vehicles that will meet Tier 2 tailpipe standards. For light-duty vehicles, these standards would not begin until model year 2004 vehicles.

Travel-demand model—an analysis procedure using heuristics or formal systems of equations to estimate the number, distribution, mode choice, and/or route choice of trips made by a household or individual that can be aggregated to estimate the number of trips starting and/or ending in a specific geographic area. The model determines the amount of transportation activity occurring in a region based on an understanding of the daily activities of individuals and employers as well as the resources and transportation infrastructure available to households and individuals when making their daily activity and travel decisions.

Transportation Control Measure (TCM)—Any control measure to reduce vehicle trips, vehicle use, vehicle miles traveled, vehicle idling, or traffic congestion for the purpose of reducing motor vehicle emissions. TCMs can include encouraging the use of carpools and mass transit.

Transportation Improvement Program (TIP)—Also known as a transportation program, a TIP is a short-term prioritized list of projects (covering 3 years out at a minimum and updated at least every 2 years) proposed to be funded or approved by the Federal Highways Administration or Federal Transit Authority. The TIP-listed projects are drawn from or consistent with the long-range transportation plan.

Transportation Plan— A long-range plan that identifies facilities that should function as an integrated transportation system. Under the Intermodal Surface Transportation Efficiency Act of 1991, MPOs must have transportation plans in place that present a 20-year perspective on transportation investments for their region. The transportation plan gives emphasis to those facilities that serve important national and regional transportation functions, and includes a financial plan that demonstrates how the long-range plan can be implemented.

Urban Airshed Model (UAM)— An air quality model that is a three-dimensional photochemical grid model calculating the concentrations of both inert and chemically reactive pollutants in the atmosphere. It simulates the physical and chemical processes that affect pollution concentrations.

Vehicle miles traveled (VMT)— The number of miles driven by a single vehicle, or by a fleet of vehicles over a set period of time, such as a day, month, or year.

Volatile organic compounds (VOCs)— An organic compound is a compound containing carbon combined with atoms of other elements, commonly hydrogen, oxygen, and nitrogen. Simple carbon-containing compounds such as carbon monoxide and carbon dioxide are usually classified as inorganic compounds. A volatile organic compound is a compound that can exist as a gas under typical atmospheric conditions. Many volatile organic chemicals are hazardous air pollutants; for example, benzene causes cancer. See Appendix B for details of how the term VOC is used in this report.

Zero Emission Vehicle (ZEV)— Vehicles which produce no emissions from the on-board source of power, e.g., an electric vehicle. Sources: California Air Resources Board at www.arb.ca.gov/html/gloss. htm; U.S. EPA at www.epa.gov/oar/oaqps/peg_caa/pegcaa10.html; U.S. EPA at www.epa.gov/oms/stds-ld.htm; Harvey and Deakin 1993; Davis 1997; FHWA 1997.

Appendix A

BIOGRAPHICAL INFORMATION ON THE COMMITTEE TO REVIEW EPA's MOBILE SOURCE EMISSIONS FACTOR (MOBILE) MODEL

Armistead G. Russell (Chair) is the Georgia Power Professor of Civil and Environmental Engineering at the Georgia Institute of Technology. He received a B.S. from Washington State University and an M.S. and Ph.D. in mechanical engineering from the California Institute of Technology. His research areas include air pollution control, aerosol dynamics, atmospheric chemistry, emissions control, air pollution control strategy design and computer modeling.

Matthew J. Barth is an associate professor of electrical engineering and the manager of transportation systems research at the Center for Environmental Research and Technology of the College of Engineering at the University of California, Riverside. He received his B.S. from the University of Colorado and an M.S. and Ph.D. in electrical and computer engineering from the University of California, Santa Barbara.

John C. Bailar III is a professor in the Department of Health Studies at the University of Chicago. His research interests include research administration, biometrics-biostatistics, public health and epidemiology, and science policy. He earned a B.A. in chemistry from the University of Colorado, an M.D. at Yale University, and a Ph.D. in statistics at American University.

Laurence S.Caretto is dean of the College of Engineering and Computer Science at California State University, Northridge (CSUN). He received his Ph.D. in engineering from the University of California, Los Angeles.

Carleton J.Howard is a research chemist for the Aeronomy Lab of the National Oceanic and Atmospheric Administration in Boulder, Colorado. He received his Ph.D. in physical chemistry from the University of Pittsburgh. He conducts research in atmospheric chemistry including laboratory studies of the kinetics of the gas reactions of atoms, radicals, and ions with small molecules.

John H.Johnson is a presidential professor of mechanical engineering at Michigan Technological University. He received his Ph.D. from the University of Wisconsin-Madison and is a fellow member of the Society of Automotive Engineers. His research involves developing diesel participate emissions measurement and control systems, methods for determining the effectiveness for diesel particulate and gaseous emissions control systems in underground mines, and a vehicle engine cooling system simulation model.

John F.Kowalczyk recently retired as manager of the air quality planning division of the Oregon Department of Air Quality. He received his M.S. in environmental engineering from Oregon State University.

Alan C.Lloyd is chairman of the California Air Resources Board. Previously, he served as executive director of the Energy and Environmental Engineering Center at the Desert Research Institute. His research interests involve alternative fuels, renewable energy and advanced technologies. He received his Ph.D. in gas kinetics from University College of Wales, Aberystwyth.

Michael R.Morris is director of transportation for the North Central Texas Council of Governments. He is responsible for providing travel input data to the MOBILE model as well as using the MOBILE model to meet EPA's transportation conformity requirements. He holds a master's degree in civil engineering from the State University of New York at Buffalo and is a licensed professional engineer.

Alison K.Pollack is a principal at ENVIRON Corporation, an environmental consulting firm. She received her M.S. in statistics from the University of Wisconsin-Madison. Her work is in the analysis of mobile source emissions and emissions models, mobile source control program evalua tion, emissions and air quality evaluation of alternative and reformulated fuels, and environmental statistics.

ROBERT F.SAWYER is a professor emeritus with the Department of Mechanical Engineering at the University of California, Berkeley. He conducts research in engine combustion, pollutant formation and control, toxic waster incineration, and alternative fuels. He received his Ph.D. from the Department of Aerospace Sciences at Princeton University.

Appendix B

ACRONYMS AND NAMES USED FOR CLASSIFYING ORGANIC COMPOUNDS

Common Abbreviation	*Full Name*	*Definition*
VOC[1]	Volatile organic compound	Organic compounds that are found in the gas phase at ambient conditions. Might not include methane.
ROG	Reactive organic gas	Organic compounds that are assumed to be reactive at urban (and possibly regional) scales. Definitionally, taken as those organic compounds that are regulated because they lead to ozone formation. Does not include methane. The term is predominantly used in California.
NMHC	Nonmethane hydrocarbon	All hydrocarbons except methane; sometimes used to denote ROG.
NMOC	Nonmethane organic compound	Organic compounds other than methane

Common Abbreviation	*Full Name*	*Definition*
RHC	Reactive hydrocarbon	All reactive hydrocarbons; also used to denote ROG.
THC	Total hydrocarbon	All hydrocarbons, sometimes used to denote VOC.
OMHCE	Organic material hydrocarbon equivalent	Organic compound mass minus oxygen mass.
TOG	Total organic gas	Used interchangeably with VOC.

[1]Unless noted otherwise, VOC's is the term used in this report to represent the general class of gaseous organic compounds.
Source: NRC 1999.